AUTOPOIESIS AND COGNITION

BOSTON STUDIES IN THE PHILOSOPHY OF SCIENCE

EDITED BY ROBERT S. COHEN AND MARX W. WARTOFSKY

VOLUME 42

HUMBERTO R. MATURANA and FRANCISCO J. VARELA

AUTOPOIESIS AND COGNITION

The Realization of the Living

With a preface to 'Autopoiesis'

by

Stafford Beer

D. REIDEL PUBLISHING COMPANY

DORDRECHT : HOLLAND / BOSTON : U.S.A.

LONDON : ENGLAND

Library of Congress Cataloging in Publication Data

CIP

Maturana, Humberto R. 1928–
Autopoiesis and cognition.

(Boston studies in the philosophy of science ; v. 42)
Bibliography: p.
Includes indexes.
1. Biology–Philosophy. 2. Cognition. 3. Life (Biology)
I. Varela, Francisco J., 1945– joint author. II. Title. III. Series.
Q174.B67 vol. 42 [QH331] 501s [574'.01] 79–24724

ISBN 90–277–1015–5
ISBN 90–277–1016–3 pbk.

Published by D. Reidel Publishing Company,
P.O. Box 17, 3300 AA Dordrecht, Holland

Sold and distributed in the U.S.A. and Canada
by Kluwer Boston Inc., Lincoln Building,
160 Old Derby Street, Hingham, MA 02043, U.S.A.

In all other countries, sold and distributed
by Kluwer Academic Publishers Group,
P.O. Box 322, 3300 AH Dordrecht, Holland

D. Reidel Publishing Company is a member of the Kluwer Group

3-0985-500 ts

Autopoiesis: The Organization of the Living was originally published in Chile
under the title *De Maquinas y Seres Vivos,* © 1972 by
Editorial Universitaria S.A.

Printed in The Netherlands

EDITORIAL PREFACE

This is a bold, brilliant, provocative and puzzling work. It demands a radical shift in standpoint, an almost paradoxical posture in which living systems are described in terms of what lies outside the domain of descriptions. Professor Humberto Maturana, with his colleague Francisco Varela, have undertaken the construction of a systematic theoretical biology which attempts to define living systems *not* as they are objects of observation and description, nor even as interacting systems, but as self-contained unities whose only reference is to themselves. Thus, the standpoint of description of such unities from the 'outside', i.e., by an observer, already seems to violate the fundamental requirement which Maturana and Varela posit for the characterization of such systems – namely, that they are autonomous, self-referring and self-constructing closed systems – in short, *autopoietic systems* in their terms. Yet, on the basis of such a conceptual method, and such a theory of living systems, Maturana goes on to define cognition as a biological phenomenon; as, in effect, the very nature of all living systems. And on this basis, to generate the very domains of interaction among such systems which constitute language, description and thinking.

The radical shift in standpoint here requires an imaginative leap and the abandonment at the outset of the standard characterizations of living systems in terms of function or purpose, or of organism-environment relations, or of causal interactions with an external world, or even in terms of information, coding and transmission. In effect, Maturana and Varela propose a theoretical biology which is topological, and a topology in which elements and their relations constitute a closed system, or more radically still, one which from the 'point of view' of the system itself, is entirely self-referential and has no 'outside', Leibnizian for our day.

The work demands and deserves careful reading. It is technical, formal, difficult, philosophical and boldly imaginative. It is rigorously constructed, and insofar as it is a theoretical biology, it remains uncompromisingly abstract and formal. Yet it smells of the medical laboratory and of the working domain of the neurophysiologist. Where the interpretation of the formal theory maps it into the domain of the nervous system, the insights and suggestions for further interpretation are exciting indeed. And we expect nothing less, here and to come.

Maturana is perhaps best known to the philosopher and the scientific nonspecialist as a co-author of the classic 1959 paper 'What the Frog's Eye Tells the Frog's Brain' (with Lettvin, McCulloch and Pitts). Since then, he has worked on the anatomy and neurophysiology of vision, especially on color vision. He has also been teaching medical students. The problems and puzzles which emerged in his research and teaching led Maturana to develop a distinctively alternative theoretical framework in order to answer the questions, 'What is a Living System?' and 'What is Cognition?' The consequence of his investigations, and of his construction of living systems as self-making, self-referring autonomous unities, is that he discovered that the two questions have a common answer. He writes, "Living systems are cognitive systems, and living as a process is a process of cognition."

We are very pleased to introduce this major theoretical work in the *Boston Studies in the Philosophy of Science*. The integration of biological theory, formal construction, epistemology (and, further, Maturana's suggestions of the nature of interacting systems as a kind of biological sociology, and his sketch of the ethical implications of such a construction) – all mark these two studies as among the most original attempts at a systematic biology in decades, and as a profoundly philosophical work.

Center for Philosophy and History of Science
Boston University
July, 1979.

ROBERT S. COHEN
MARX W. WARTOFSKY

GENERAL TABLE OF CONTENTS

BIOLOGY OF COGNITION

AUTOPOIESIS: THE ORGANIZATION OF THE LIVING

FOREWORD

Everything that I say in the introduction is my exclusive responsibility. Francisco Varela has been generous enough to grant me this freedom in what concerns the essay that we wrote together. His views about it he expresses fully and independently in his book *Principles of Biological Autonomy*, published by Elsevier-North Holland, New York, 1979.

Also, I wish to acknowledge the subtle debt which the contents of this book owe to the many illuminating conversations about all the topics here contained that I have had over more than fourteen years with my dear friend Professor Felix Schwartzmann, of the University of Chile.

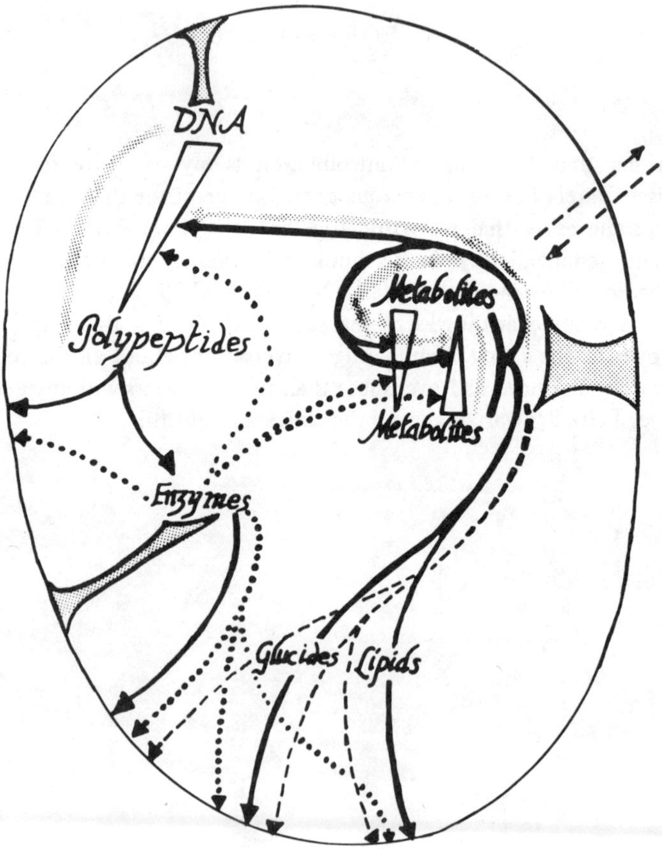

Representation of the cellular autopoietic network.

All arrows that do not cross the boundary of the represented unity indicate production relations. The uniformly shaded areas, including the boundary line and the wedges, together with the names, indicate constitutive relations. The general form of closure with respect to production and constitution in a recursive network realized as a concrete unity through the preferential relations of the components within the network, indicate order relations and the consequent cleavage of the network as a simple unity from its medium. The whole represents a closed network of productions, but the arrows across the depicted constitutive boundary of the network indicate the necessary material openness of the system as it realizes the physical space.

INTRODUCTION

In this introduction I wish to make some comments in relation to the two essays that make this book, but in order to do so I must, at least in part, write about how they came into being.

HISTORY

Since my childhood I have been interested in animals and plants, and I frequently asked myself what made them living. Thus, in 1948, in my first year as a medical student, I wrote a poem whose first stanza was:

"¿Qué es la muerte para el que la mira?
¿Qué es la muerte para el que la siente?
Pesadez ignota, incomprensible,
dolor que el egoísmo trae, para ése;
silencio, paz y nada, para éste.
Sin embargo el uno siente
que su orgullo se rebela, que su mente
no soporta que tras la muerte nada quede,
que tras la muerte esté la muerte.
El otro, en su paz, en su silencio,
en su majestad inconsciente siente,
nada siente, nada sabe,
porque la muerte es la muerte
y tras la muerte está la vida
que sin la muerte sólo es muerte."

What is death for the beholder? / What is death for the dying? / A weight beyond knowledge or understanding, / A pain for the self-asserting ego, for the one; / For the other, silence, peace, and nothingness. // Yet the one feels his pride in anger / And in his mind he does not accept / That beyond death nothing should arise, / And that beyond death / There should be only death. // The other, in his silence, / In his unknowing majesty feels, / He feels nothing, he knows nothing, / Because death is death / And life without death is only emptiness. //

The poem is not a very good one, yet it contains the implicit question: 'What kind of systems are living systems that they may die, and how come that they cognize?'

In 1954 I went to study biology in England and the U.S.A., and when I returned to Chile in 1960, after six years of study and research abroad, I began to work at the Medical School of the University of Chile in Santiago as a research associate in the Department of Biology. There I was involved in two kinds of activities; I collaborated in teaching a course in general biology for the medical students, and I did research in the fields of neurophysiology and neuroanatomy. In my teaching my responsibility was to convey to the medical students some understanding of the organization of living systems as autonomous entities, as well as some understanding of their possible origin on earth. In my research I wanted to apply to the study of form and color vision in birds the same approach that J. Y. Lettvin and I had used in the study of form vision in the frog.

I soon discovered through my teaching that the central question that the students would always ask was: 'What is proper to living systems that had its origin when they originated, and has remained invariant since then in the succession of their generations?' At the same time I soon realized in my research that my central purpose in the study of color vision could not be the study of a mapping of a colorful world on the nervous system, but rather that it had to be the understanding of the participation of the retina (or nervous system) in the generation of the color space of the observer.

As a result of these different activities I entered a situation in which my academic life was divided, and I oriented myself in search of the answers to two questions that seemed to lead in opposite directions, namely: 'What is the organization of the living?' and 'What takes place in the phenomenon of perception?'

Let me speak about how I faced them.

First Question: What is the organization of the living?

When this question was first asked by the students, although it was the same question that had been lurking in my mind for many years, I could not answer it. I had prepared myself for this moment, but when it came and I tried to answer it in a manner satisfactory for the students I realized that I had to think everything anew. I could speak about form and function and astonish my students and myself at the harmony of nature, exalting the fitness of the environment and the fitness of the individual. I could claim that the question was a very difficult one and that it could not yet be answered due to our insufficient knowledge. We had to accept that we could recognize living systems when we encountered them, but that we could not yet say what they were. I could enumerate features of living systems such as reproduction, heredity, growth, irritability, and so on; but, how long a list was necessary?; when would the list be completed? In order to know when the list was completed I had to know what a living system was, which was, in fact, the question that I wanted to answer in the first place by producing such a list. I could speak about adaptation and evolution, about development and differentiation, and show how all these phenomena were tied together by the phenomenon of natural selection; but the question: 'What was the invariant feature of living systems around which natural selection operated?', remained unanswered. Every approach that I could attempt and that I did attempt left me at the starting point.

Yet I obviously had some inkling of what was the correct answere, because I rejected the unsatisfactory ones. After several years of these various attempts I realized that the difficulty was both epistemological and linguistic, and that both my wife and my old professor, J. Z. Young, were right: one can only say with a given language what the language permits. I had to stop looking at living systems as open systems defined in an environment, and I needed a language that would permit me to describe an autonomous system in a manner that retained autonomy as a feature of the system or entity specified by the description. In other words, any attempt to characterize living systems with notions of purpose or function was doomed to fail because these notions are intrinsically referential and cannot be operationally used to characterize any system as an autonomous entity. Therefore, notions of purpose, goal, use or function, had to be rejected, but initially I did not know how. Accordingly, I tried in my lectures several approaches in order to find a way of speaking about living systems in a manner that would grasp their autonomy as a phenomenon of their operation as unitary systems. Thus, eventually, I made a distinction between what I called self-referred and allo-referred systems, a distinction that separated systems that could only be characterized with reference to themselves, such as living systems, from systems that could only be characterized with reference to a context. I did this in order to emphasize that whatever took place in living systems as living systems, took place as necessarily and constitutively determined in relation to themselves because their being defined as unities through self-reference was their manner of autonomy; and that whatever took place in other systems took place as constitutively determined in relation to the context with respect to which they were defined as unities. This way of speaking was not fully satisfactory but led me to realize that what was indeed needed was the characterization of a kind of system which, if allowed to operate, would operate in a manner indistinguishable from the operation of living systems, and that one should do so using only neighborhood relations realized through the properties of the components of the system. It was with such aim that I spoke for the first time in 1969 of living systems as systems defined as unities through the basic circularity of their production of their components.

Second question: What takes place in the phenomenon of perception?

When Jerry Y. Lettvin and I wrote our several articles on frog vision [Lettvin, J. Y., H. R. Maturana, W. S. McCulloch and W. H. Pitts 1959; Maturana, H. R., J. Y. Lettvin, W. S. McCulloch and W. H. Pitts 1960], we did it with the implicit assumption that we were handling a clearly defined cognitive situation: there was an objective (absolute) reality, external to the animal, and independent of it (not determined by it), which it could perceive (cognize), and the animal could use the information obtained in its perception to compute a behavior adequate to the perceived situation. This assumption of ours appeared clearly in our language. We described the various kinds of retinal ganglion cells as feature detectors, and we spoke about the detection of prey and enemy. We knew that was not the whole neurophysiological story, as was apparent particularly in the discussion of the article called 'Anatomy and Physiology of Vision in the Frog (*Rana pipiens*)'. But even there the epistemology that guided our thinking and writing was that of an objective reality independent of the observer. Thus, when Samy Frenk and I began to work with pigeons in 1961, first studying form vision, we approached that study with the same fundamental view. No problem arose then and without any difficulty we could characterize many classes of retinal ganglion cells. Yet, when Gabriela Uribe joined us and we in fact began to study color vision in 1964, it soon became apparent to us that that approach leads to deep trouble. Neurophysiologically we did not see anything fundamentally different from what other scholars had already seen. We found the classic types of ganglion cells with separate, concentric or overlapping opponent spectral preferences. But we also found: (a) that although the geometry of the receptive fields of the ganglion cells with opponent spectral preferences had nothing to do with the geometry of the visual object, the geometry of the visual object had to do with the response of those cells; and (b) that we could not account for the manifold chromatic experiences of the observer by mapping the visible colorful world upon the activity of the nervous system, because the nervous system seemed to use geometric relations to specify color distinctions. A different approach and a different epistemology were necessary.

There are many visual configurations, with uniform and variegated spectral compositions, in simple and complex geometrical forms, that give rise to indistinguishable color experiences. How should one, then, look for the invariances in the activity of the nervous system, if any, in relation to the perception of color? After we realized that the mapping of the external world was an inadequate approach, we found that the very formulation of the question gave us the clue. What if, instead of attempting to correlate the activity in the retina with the physical stimuli external to the organism, we did otherwise, and tried to correlate the activity in the retina with the color experience of the subject?

Such a step entailed two difficulties. On the one hand it required the definition of a reference which would permit the characterization of the activity of the retinal ganglion cells independently of the stimulus as such; on the other hand it required us to close the nervous system and treat the report of the color experience as if it represented the state of the nervous system as a whole. In other words, the new approach required us to treat seriously the activity of the nervous system as determined by the nervous system itself, and not by the external world; thus the external world would only have a triggering role in the release of the internally-determined activity of the nervous system. We did this rigorously, and showed that such an approach did indeed permit us to generate the whole color space of the observer. That was a very fundamental result that we published in a very unknown article [Maturana, H. R., G. Uribe, and S. Frenk, 1968].

But what was still more fundamental was the discovery that one had to close off the nervous system to account for its operation, and that perception should not be viewed as a grasping of an external reality, but rather as the specification of one, because no distinction was possible between perception and hallucination in the operation of the nervous system as a closed network. Although we arrived at this conclusion through the study of color vision, there are many earlier experimental studies (such as those of Stone on the rotation of the eye of the salamander in the early 'forties) that could also have led to an understanding of the nervous system as a closed network of interacting neurons. Whether they did or not, I do not know; but if they did it seems that the implications were not pursued to their ultimate consequences.

Whatever the case, for me this finding had great significance and plunged me into the study of cognition as a legitimate biological problem. Two inmediate consequences arose from this: the first one was that in my neurophysiological studies I had to take seriously the indistinguishability in the operation of the nervous system between perception and hallucination; the second one was that I needed a new language to talk about the phenomena of perception and cognition. The first consequence required that the question: 'How does the organism obtain information about its environment?' be changed to: 'How does it happen that the organism has the structure that permits it to operate adequately in the medium in which it exists?' A semantic question had to be changed into a structural question. The second question required the actual attempt to describe the phenomena that take place in the organism during the occurrence of the phenomena of perception and cognition in a language that retained them as phenomena proper to a closed nervous system.

A Congress in Anthropology

Early in May of 1968 the University of Chile entered a state of revolution. The students took over the University in an attempt to reformulate the philosophy that had inspired its organization. I joined them. All standard academic activities stopped and students and some members of the faculty tried to say something new. It was not easy. Language was a trap, but the whole experience was a wonderful school in which one could discover how mute, deaf and blind one was. It was easy to be caught in one's own ego, but if one succeeded in attaining at least some degree of freedom from it, one began to listen and one's language began to change; and then, but only then, new things could be said. This lasted for several months.

In September of that year, I accepted an invitation to go to the University of Illinois at Urbana to the Biological Computer Laboratory of Professor Heinz von Foerster. Furthermore, Professor von Foerster invited me to participate in a symposium sponsored by the Wenner–Gren Foundation for Anthropological Research to be held during March 1969 in Chicago, with the purpose of considering the anthropology of cognition. The invitation was to speak on 'The state of the art of the neurophysiology of cognition'. I accepted this invitation and decided not to speak about neuronal circuits, nerve impulses or synapses, but rather I decided to consider what should take place in the organism during cognition by considering cognition as a biological phenomenon. In doing this I found that my two apparently contradictory academic activities were not contradictory, and that they were, in fact, addressed to the same phenomenon: cognition and the operation of the living system – its nervous system included when present – were the same thing. From this understanding the essay 'Biology of Cognition' arose as an expansion of my presentation in that symposium.

The Word 'Autopoiesis'

The second essay included in this book was written in 1972, as an expansion of the section on 'Living Systems' in the 'Biology of Cognition'. The writing of this essay was in fact triggered by a conversation that Francisco Varela and I had in which he said: "If indeed the circular organizaton is sufficient to characterize living systems as unities, then one should be able to put it in more formal terms". I agreed, but said that a formalization could only come after a complete linguistic description, and we immediately began to work on the complete description. Yet we were unhappy with the expression 'circular organization', and we wanted a word that would by itself convey the central feature of the organization of the living, which is autonomy. It was in these circumstances that one day, while talking with a friend (José Bulnes) about an essay of his on Don Quixote de la Mancha, in which he analyzed Don Quixote's dilemma of whether to follow the path of arms (*praxis*, action) or the path of letters (*poiesis*, creation, production), and his eventual choice of the path of *praxis* deferring any attempt at *poiesis*, I understood for the first time the power of the word 'poiesis' and invented the word that we needed: *autopoiesis*. This was a word without a history, a word that could directly mean what takes place in the dynamics of the autonomy proper to living systems. Curiously, but not surprisingly, the invention of this word proved of great value. It simplified enormously the task of talking about the organization of the living without falling into the always gaping trap of not saying anything new because the language does not permit it. We could not escape being immersed in a tradition, but with an adequate language we could orient ourselves differently and, perhaps, from the new perspective generate a new tradition.

Let me now say something about the essays themselves.

BIOLOGY OF COGNITION

When I wrote this essay I did not yet have the word 'autopoiesis', nor had I come to the more formal expression of the living organization given in the next essay. Yet, these shortcomings do not detract from what is said because the basic relations embodied in the notion of autopoiesis are fully implied, although cumbersomely said, by expressions such as 'circular organization' and 'self-referential systems'. Furthermore, when I wrote the essay I decided not to make any concession to existing notions that I considered inadequate or misleading, even if this seemed to make the text particularly obscure. However, I made a concession which I have always regretted. I submitted to the pressure of my friends and talked about causal relations when speaking about the circular organization of living systems. To do this was both inadequate and misleading. It was inadequate because the notion of causality is a notion that pertains to the domain of descriptions, and as such it is relevant only in the metadomain in which the observer makes his commentaries and cannot be deemed to be operative in the phenomenal domain, the object of the description. It was misleading because it obscured the actual appreciation of the sufficiency of the notion of property as defined by the distinctive operation performed by the observer when specifying a unity, for the description of the phenomenal domains generated by the specified unities. It was misleading because it obscured the understanding of the dependency of the identity of the unity on the distinctive operation that specified it. It was misleading because it obscured both the understanding of the phenomenal domains as determined by the properties of the unities that generate them, and the non-intersection of the phenomenal domains generated by the operation of a composite unity as a simple unity in a medium and by the operation of its components as components.

There is nothing else that I wish to add as a commentary to this essay. It is a cosmology and as such it is complete. Finally I wish to say that I find it pervading my views and understanding of everything. In a sense it has been my way to transcendental experience: to the discovery that matter, metaphorically speaking, is the creation of the spirit (the mode of existence of the observer in a domain of discourse), and that the spirit is the creation of the matter it creates. This is not a paradox, but it is the expression of our existence in a domain of cognition in which the content of cognition is cognition itself. Beyond that nothing can be said.

AUTOPOIESIS

This article was written as an expansion of two pages of the 'Biology of Cognition', in an attempt to show that autopoiesis is necessary and sufficient to characterize the organization of living systems, and that given the proper historical contingencies one can derive all the biological phenomenology from the characterization of living systems as autopoietic systems in the physical space. Notions of purpose, function or goal are unnecessary and misleading. This the essay does, and in this respect no commentary is needed; the essay stands by itself. Yet, when we wrote it we were just beginning to realize the fundamental distinction between organization and structure, and we do not separate the terms with complete rigor. Also, because it was not written under the supposition that the reader had read the 'Biology of Cognition', the essay is not fully clear concerning the validity of the statement "Everything said is said by an observer" in relation to the distinction between characterization and description of a system. I shall make some comments about these themes.

Unity, Organization and Structure

Unity. The basic cognitive operation that we perform as observers is the operation of distinction. By means of this operation we specify a unity as an entity distinct from a background, characterize both unity and background with the properties with which this operation endows them, and specify their separability. A unity thus specified is a simple unity that defines through its properties the space in which it exists and the phenomenal domain which it may generate in its interactions with other unities.

If we recursively apply the operation of distinction to a unity, so that we distinguish components in it, we respecify it as a composite unity that exists in the space that its components define because it is through the specified properties of its components that we observers distinguish it. Yet we can always treat a composite unity as a simple unity that does not exist in the space of its components, but which exists in a space that it defines through the properties that characterize it as a simple unity. Accordingly, if an autopoietic system is treated as a composite unity, it exists in the space defined by its components, but if it is treated as a simple unity the distinctions that specify it as a simple unity characterize its properties as a simple unity, and define the space in which it exists as such a simple unity.

Organization and Structure. The relations between components that define a composite unity (system) as a composite unity of a particular kind, constitute its organization. In this definition of organization the components are viewed only in relation to their participation in the constitution of the unity (whole) that they integrate. This is why nothing is said in it about the properties that the components of a particular unity may have other than those required by the realization of the organization of the unity.

The actual components (all their properties included) and the actual relations holding between them that concretely realize a system as a particular member of the class (kind) of composite unities to which it belongs by its organization, constitute its structure. Therefore, the organization of a system as the set of relations between its components that define it as a system of a particular class, is a subset of the relations included in its structure. It follows that any given organization may be realized through many different structures, and that different subsets of relations included in the structure of a given entity, may be abstracted by an observer (or its operational equivalent) as organizations that define different classes of composite unities.

The organization of a system, then, specifies the class identity of a system, and must remain invariant for the class identity of the system to remain invariant: if the organization of a system changes, then its identity changes and it becomes a unity of another kind. Yet, since a particular organization can be realized by systems with otherwise different structures, the identity of a system may stay invariant while its structure changes within limits determined by its organization. If these limits are overstepped, that is, if the structure of the system changes so that its organization cannot any more be realized, the system loses its identity and the entity becomes something else, a unity defined by another organization.

It is apparent that only a composite unity has structure and organization, a simple unity does not. A simple unity only has the properties with which it is endowed by the operations of distinction through which it becomes separated from a background. It is also apparent that as soon as a composite unity is treated as a simple unity, any question about the origin of its properties becomes inadequate because the properties of a simple unity are given through its distinction as a simple unity. It is also apparent that although the properties of a composite unity as a simple unity arise from its organization, they are realized through the properties of its components. Accordingly, while two simple unities interact through the simple interplay of their properties, two composite unities interact in a manner determined by their organization and structure through the interplay of the properties of their components.

Structural Coupling

In the history of interactions of a composite unity in its medium, both unity and medium operate in each interaction as independent systems that, by triggering in each other a structural change, select in each other a structural change. If the organization of a composite unity remains invariant while it undergoes structural changes triggered and selected through its recurrent interactions in its medium, that is, its adaptation is conserved, then the outcome of this history of interactions is the selection, by the recurrent or changing structural configuration of the medium, of a sequence of structural changes in the composite unity, which results in that the changing structure of the organism follows the changing structure of the medium through a continued structural coupling to it. If organization or adaptation are not conserved, then the outcome for the composite unity is disintegration. In other words, if a composite unity is structurally plastic its conservation of adaptation results in its maintained structural coupling to the medium that selects its path of structural change. In this process, the configuration of constitutive relations that remain invariant in the adapted composite unity, determines the matrix of possible perturbations that the composite unity admits at any instant, and, hence, operates as a reference for the selection of the path of structural changes that take place in it in its history of interactions. Defined in this manner, structural coupling (conservation of adaptation) is not peculiar to living systems. It is a phenomenon that takes place whenever a plastic composite unity undergoes recurrent interactions with structural change but without loss of organization, which may follow any changing or recurrent structural configuration of its domain of interactions (medium). Therefore, all that is unique with respect to adaptation in living systems is that in them the autopoietic organization constitutes the invariant configuration of relations around which the selection of their structural changes takes place during their history of interactions.

Epistemology

As soon as a unity is specified, a phenomenal domain is defined. Accordingly, if a composite unity operates as a simple unity, it operates in a phenomenal domain that it defines as a simple unity that is necessarily different from the phenomenal domain in which its components operate. Therefore, the emergence of a phenomenal domain as the result of the operational distinction of a composite unity as a simple unity, makes phenomenal reductionism (and, hence, explanatory reductionism) impossible. Furthermore, the dynamics of the establishment of unities through operational distinctions that specify their properties, have the result that all phenomenal domains are necessarily realized through the operation (interplay) of the properties of the unities that generate them, that is, through relations of contiguity. If a component '*A*' through its interaction with a component '*B*' triggers an interaction of '*B*' with '*C*' that triggers a reduction in the production of '*D*', an observer may say by considering the whole that '*A*' controls the production of '*D*'. '*A*', '*B*', '*C*' and '*D*', interact through relations of contiguity in the phenomenal domain that the components define. Relations such as regulation, control or function, therefore, are not relations of contiguity; they are referential relations specified by the observer who puts himself in a metadomain of descriptions by using his view of the whole as a reference for his description of the participation of the components that he describes in the constitution of the composite unity.

"Everything said is said by an observer." This I say in the 'Biology of Cognition'. The fundamental cognitive operation that an observer performs is the operation of distinction. By means of this operation the observer specifies a unity as an entity distinct from a background and a background as the domain in which an entity is distinguished. An operation of distinction, however, is also a prescription of a procedure which, if carried out, severs a unity from a background, regardless of the procedure of distinction and regardless of whether the procedure is carried out by an observer or by another entity. Furthermore, the prescriptiveness of an operation of distinction implies a universal phenomenalism of distinctions which, through the specification of new procedures of distinction or through their recursive application in the reordering of the distinguished entities, can, in principle, endlessly give rise to new simple and composite unities, and, hence, to new non-intersecting phenomenal domains. Thus, although a distinction performed by an observer is a cognitive distinction and, strictly, the unity thus specified exists in his cognitive domain as a description, the observer in his discourse specifies a metadomain of descriptions from the perspective of which he establishes a reference that allows him to speak as if a unity, simple or composite, existed as a separate entity that he can characterize by denoting or connoting the operations that must be performed to distinguish it.

In the perspective of a descriptive metadomain the distinction between the characterization of a unity and the observer's knowledge of it that permits him to describe it in a context, should be clear. In fact, knowledge always implies a concrete or conceptual action in some domain, and the recognition of knowledge always implies an observer that beholds the action from a metadomain. Therefore, when an observer claims knowledge of a system, he claims that he can define a metadomain from the perspective of which he can simultaneously behold the system as a simple unity, describing its interactions and relations as a simple unity, and its components as components, describing their interactions and relations as components. In these circumstances it is legitimate to distinguish between the characterization that an observer makes of a unity, either by pointing to its properties if it is a simple unity, or by pointing to its organization if it is a composite one, from the knowledge about a unity that he reveals, either by describing its operation as a simple unity if it is a simple unity, or by describing both its operation as a simple unity and the operation of its components if it is a composite entity. In either case, however, the knowledge that an observer claims of the unities that he distinguishes consists in his handling of them in a metadomain of descriptions with respect to the domain in which he characterizes them. Or, in other words, an observer characterizes a unity by stating the conditions in which it exists as a distinguishable entity, but he cognizes it only to the extent that he defines a metadomain in which he can operate with the entity that he characterized.

Thus, autopoiesis in the physical space characterizes living systems because it determines the distinctions that we can perform in our interactions when we specify them, but we know them only as long as we can both operate with their internal dynamics of states as composite unities and interact with them as simple unities in the environment in which we behold them. The fact that the characterization of an entity is also a description made by the observer, and as such also belongs to his descriptive domain ('Biology of Cognition'), does not invalidate the operational effectiveness of the distinctions upon distinctions that constitutes the metadomain of descriptions in which the cognitive statements are made. The entity characterized is a cognitive entity, but once it is characterized the characterization is also subject to cognitive distinctions valid in the metadomain in which they are made by treating the characterization as an independent entity subject to contextual descriptions. Therefore, the complementarities system/environment, autonomy/control, totality/composition, etc., are complementarities only in our cognition of a system that we observe in a context that allows us to establish such relations, but they are not constitutive features of the referred system because they do not participate in its constitution through the interplay of the properties of its components. Accordingly, that one should not be able to account for or deduce all actual biological phenomena from the notion of autopoiesis without resorting to historical contingencies, is not a shortcoming of such a notion. On the contrary, it is to be expected because the notion of autopoiesis only characterizes living systems as autonomous entities that can be distinguished as composite unities realized through neighborhood relations.

Finally, I would like to add some sociological and ethical comments that I consider follow from the understanding of the autopoietic organization of living systems. The essay on autopoiesis was supposed to have a second appendix on social and ethical implications. This appendix, however, was never included because Francisco Varela and I never agreed about its contents. Now I shall use the privilege that I have by writing this introduction in order to present the notions that I would have included in that appendix.

SOCIETY AND ETHICS

The central feature of human existence is its occurrence in a linguistic cognitive domain. This domain is constitutively social. Yet, what is a social system? how is a social system characterized? how do living systems in general, and human beings in particular, participate in the constitution of the social system that they integrate? The answers to these questions are central for the understanding of social dynamics and the process of social change. The following considerations state the essence of my answers to these questions:

(1) It is apparent that natural social systems as systems constituted by living systems require these for their actual realization. What is not apparent, however, is the extent to which the coupling of living systems in the integration of a social system entails the realization of their autopoiesis. If the realization of a social system were to entail the autopoiesis of its components, then the realization of the autopoiesis of the components of a social system would be intrinsically indispensable and constitutive of its operation as such, and not a mere contingency. Accordingly, any particular feature, cultural or otherwise, of the manner in which the autopoiesis of the components is realized and their individuality and autonomy is restricted, would be a feature of the peculiar social system (society) in question, and not intrinsic to it as a social system. If, however, the autopoiesis of the components of a natural social system were not involved in its constitution because the relations that define a system as social do not entail them, then the autopoiesis of the components (and hence their autonomy and individuality) would be intrinsically dispensable.

(2) The question, 'What is a social system?' cannot be answered by simply describing a particular one because we do not know the significant relations that we must abstract when characterizing its organization. The question must be answered by proposing a system which, if allowed to operate, would generate a phenomenal domain indistinguishable from the phenomenal domain proper to a natural social system. Accordingly, I propose that a collection of autopoietic systems that, through the realization of their autopoiesis, interact with each other constituting and integrating a system that operates as the (or as a) medium in which they realize their autopoiesis, is indistinguishable from a natural social system. Or, in other words, I propose

that the relations stated above characterize the organization of a social system as a system, and that all the phenomena proper to social systems arise from this organization. If one accepts this proposition, and I shall henceforth proceed as if it were accepted, then one has to accept the following implications:

(i) The realization of the autopoiesis of the components of a social system is constitutive to the realization of the social system itself. This cannot be ignored in any consideration about the operation of a social system without negating it.

(ii) A collection of living systems integrating a composite unity through relations that do not involve their autopoiesis is not a social system, and the phenomena proper to its operation as such a composite unity are not social phenomena. Since the operational coupling of an organism integrating a composite unity does not necessarily involve all its properties, an observer may see an organism as integrating simultaneously several composite unities that may or may not all be social systems.

(iii) The structure of a society as a particular social system is determined both by the structure of its autopoietic components and by the actual relations that hold between them while they integrate it. Therefore, the domain of social phenomena, defined as the domain of the interactions and the relations that an observer sees taking place between the components of a society, results from the autopoietic operation of the components of the society while they realize it in the interplay of their properties.

(iv) In a society, at any instance of observation, the structures of the components determine the properties of the components, the properties of the components realize the structure of the society, and the structure of the society operates as a selector of the structure of its components by being a medium in which they realize their ontogeny.

(v) An autopoietic system participates in the constitution of a social system only to the extent that it participates in it, that is, only as it realizes the relations proper to a component of the social system. Accordingly, in principle, an autopoietic system may enter or leave a social system at any moment by just satisfying or not satisfying the proper relations, and may participate simultaneously or in succession in many different ones.

In what follows I shall pursue the consequences of these notions, and whenever I speak of a social system or of a society as a social system of a particular kind, I speak of a system defined as a system by the organization proposed above.

(3) A society defines the domain in which it is realized as a unity. Such a domain may or may not include the components of the society itself, and

may or may not include other societies, but in any case it constitutes an operationally independent medium that operates as: (a) a selector of the path of structural change that the society follows in its individual history, and (b), if stable, a historical stabilizer of the structures that realize the selected invariant relations that define the society as a particular social system. The more varied the medium of selection, the greater the domain of stabilized relations in the society and, hence, the more fixed the structure of its components. If the society is a human society this takes place in a language-centered culture, and the stabilization of the structure of the human components is realized through a cultural stabilization of the relations that they must satisfy as social entities.

(4) To the extent that human beings are autopoietic systems, all their activities as social organisms must satisfy their autopoiesis. This they do in the social domain through the fulfilment of the basic biological preferences (states of pleasure) and rejections (states of displeasure) that constitute the inmediate experiential domain in which they, as components of a society, necessarily realize their individual worlds and contribute to the determination of the individual worlds of others. In man as a social being, therefore, all actions, however individual as expressions of preferences or rejections, constitutively affect the lives of other human beings and, hence, have ethical significance.

(5) What determines the constitution of a social system are the recurrent interactions of the same autopoietic systems. In other words, any biological stabilization of the structures of the interacting organisms that results in the recurrence of their interactions, may generate a social system. Among human beings the basic stabilizing factor in the constitution of a social system is the phenomenon of love, the seeing of the other as a partner in some or all the dimensions of living. In these circumstances, when a human being makes the choice of a particular way of living, apparent in his realization of a particular set of social relations, he makes a basic ethical choice in which he validates a world for himself and for others that he has explicitly or implicitly accepted as partners in living. Accordingly, the fundamental ethical problem that a man faces as an observer-member of a society is the ethical justification of the particular relations of surrender of autonomy and individuality that he demands from himself and from other members of the society that he generates and validates with his conduct.

(6) A social system is essentially a conservative system. This is so because it is generated through the interactions of structure-determined autopoietic systems and operates as a medium that selects the path of ontogenic structural

change of its components, which, thus, become structurally coupled to it. In our case, we as social beings generate, through our structure-determined properties, our societies as the cultural media that select our individual paths of ontogenic change in a manner that leads each one of us to the structure that makes us generate the particular societies to which we belong. A society, therefore, operates as a homeostatic system that stabilizes the relations that define it as a social system of a particular kind.

(7) In general, the domain of states of a system as a composite unity is determined by the properties of its components that realize its organization. If some of these properties change because the structure of some of the components changes, then, while the system either changes its properties without change of organization or disintegrates becoming something else, the changed components either integrate the system in a different manner or uncouple from it. This also applies to social systems, including human societies, because it is the actual interplay of the properties of the components that constitutes a social system as an actual system in the space in which these exist. In these circumstances, a change in the relations that define a society as a particular social system can only take place through a change in the properties of the components that realize it. It follows that in a human society a social change can only take place if the individual properties, and, hence, conduct, of its members change.

(8) All that matters for the realization of a society is that the component autopoietic systems should satisfy certain relations regardless of the actual structures (internal processes) through which they realize them. Accordingly, hypocrisy plays an important role in the realization of human societies, permitting human beings under stress to feign having certain properties which they abandon as soon as the stress is removed. This is why in a human society a social change takes place as a permanent phenomenon only to the extent that it is a cultural change: a revolution is a revolution only if it is an ethical revolution.

(9) Interactions within a society are necessarily confirmatory of the relations that define it as a particular social system; if not, the organisms that interact do not interact as components of the society which they otherwise integrate. It is only through interactions operationally not defined within the society that a component organism can undergo interactions that lead to the selection, in its ontogeny, of a path of structural change not confirmatory of the society that it integrates. This is why social creativity, as the generation of novel social relations, always entails interactions operationally outside the society, and necessarily leads to the generation, by the creative

individuals, of modes of conduct that either change the defining relations of the society as a particular social system, or separate them from it. Social creativity is necessarily antisocial in the social domain in which it takes place.

(10) In general any organism, and in particular any human being, can be simultaneously a member of many social systems, such as a family, a club, an army, a political party, a religion or a nation, and can operate in one or another without necessarily being in internal contradiction. A human being operating as an observer, however, can always define a metadomain from the perspective of which he may see his participation in the various social systems that he integrates, and find it contradictory. Conduct as observer by a human being implies that he stands operationally as if outside the various social systems that he otherwise integrates, and that he may undergo in this manner interactions that do not confirm them. An observer always is potentially antisocial.

(11) To grow as a member of a society consists in becoming structurally coupled to it; to be structurally coupled to a society consists in having the structures that lead to the behavioral confirmation of the society. The spontaneous course of historical structural transformation of a society as a unity is toward its structural coupling to the medium in which it exists, and, therefore, toward the stabilization of the mechanisms that generate its defining relations through the stabilization of the properties of its components. In the domain of human societies this means the stabilization of human conduct. But, the stabilization of human conduct always entails a restriction of creativity through a restriction of the possible interactions of the individual human beings outside those prescribed by the society that they integrate. The extreme case of this, of course, takes place in a totalitarian society of any kind. Or, in other words, the spontaneous course of the historical transformation of a human society as a unity is towards totalitarianism; this is so because the relations that undergo historical stabilization are those that have to do with the stability of the society as a unity in a given medium, and not with the well-being of its component human beings that may operate as observers. Any other course requires an ethical choice; it would not be spontaneous, it would be a work of art, a product of human aesthetic design. If human beings were not observers, or capable of being so, the stabilization of their properties would not appear to matter because they would not be able to desire something else.

(12) We as human beings exist in a network of social systems and move from one to another in our daily activities. Yet, not all human beings caught in the mesh of relations generated in this network of social systems parti-

cipate in it as social beings. A human being that through his interactions with other human beings participates in interactions proper to their social system in a manner that does not involve his autopoiesis as a constitutive feature of it, is being used by the social system but is not one of its members. If the human being cannot escape from this situation because his life is at stake, he is under social abuse.

(13) All kinds of societies are biologically legitimate. Yet not all are equally desirable as systems in which an observer human being may wish to live. The capacity that man has as a language-centered social being of becoming an observer, and thus of operating as if he were external to the situation in which he finds himself, allows him, if he has the proper experiences, to contemplate the societies that he integrates and to like or dislike them. If the observer human being defines a metadomain from the perspective of which some of the defining relations of the society are undesirable, he dislikes it, and if he acts accordingly he becomes antisocial and may come to validate another society with his conduct. A totalitarian society restricts this possibility either by specifying the experiences that its components may have, so that they do not operate as observers, or by uncoupling the dissidents so that they may not seduce others to be observers as themselves. However, there are certain experiences that cannot be fully specified in a human society without destroying the basic individual structural plasticity needed for the establishment of consensual domains and the generation of language and, hence, for human creativity in general. Love is one of these experiences, and as long as man has a language he can become an observer through the experience of love.

(14) When a human being '*A*' encounters another human being '*B*' and loves him or her, he sees '*B*' in a social context and becomes an observer of the society that '*B*' integrates. '*A*' may like or may not like what he sees in reference to '*B*' and act accordingly, becoming antisocial if he does not like what he sees. An absolute totalitarian society must negate love as an individual experience because love, sooner or later, leads to an ethical evaluation of the society that the loved one integrates.

(15) A human society in which to see all human beings as equivalent to oneself, and to love them, is operationally legitimate without demanding from them a larger surrender of individuality and autonomy than the measure that one is willing to accept for oneself while integrating it as an observer, is a product of human art, that is, an artificial society that admits change and accepts every human being as not dispensable. Such a society is necessarily a non-hierarchical society for which all relations of order are constitutively

transitory and circumstantial to the creation of relations that continuously negate the institutionalization of human abuse. Such a society is in its essence an anarchist society, a society made for and by observers that would not surrender their condition of observers as their only claim to social freedom and mutual respect.

At this point there is either much more to say, or nothing – therefore, let us let the reader judge. Thank you.

BIOLOGY OF COGNITION
(1970)

By

HUMBERTO R. MATURANA

I dedicate this book to my wife, Maria Montañez Luna. I could not have written it if I had not lived as I have lived, and for the most part of my existence the thread of my being has been and is braided with hers. Therefore, strictly, although I wrote this book, she is, in every respect, as much its author as I am. This I wish both to acknowledge and to thank with deep love.

TABLE OF CONTENTS

HUMBERTO R. MATURANA

BIOLOGY OF COGNITION

I. INTRODUCTION

Man knows and his capacity to know depends on his biological integrity; furthermore, he knows that he knows. As a basic psychological and, hence, biological function *cognition* guides his handling of the universe and *knowledge* gives certainty to his acts; *objective knowledge* seems possible and through *objective knowledge* the universe appears systematic and predictable. Yet *knowledge* as an experience is something personal and private that cannot be transferred, and that which one believes to be transferable, *objective knowledge*, must always be created by the listener: the listener understands, and *objective knowledge* appears transferred, only if he is prepared to understand. Thus *cognition* as a biological function is such that the answer to the question, '*What is cognition*?' must arise from understanding knowledge and the knower through the latter's capacity to know.

Such is my endeavor.

Epistemology

The basic claim of science is objectivity: it attempts, through the application of a well defined methodology, to make statements about the universe. At the very root of this claim, however, lies its weakness: the *a priori* assumption that objective knowledge constitutes a description of that which is known. Such assumption begs the questions, '*What is it to know?*' and '*How do we know?*'.

Biology

(a) The greatest hindrance in the understanding of the living organization lies in the impossibility of accounting for it by the enumeration of its properties; it must be understood as a unity. But if the organism is a unity, in what sense are its component properties it parts? The organismic approach does not answer this question, it merely restates it by insisting that there are elements of organization that subordinate each part to the whole and make the organism a unity [Cf. Bertalanffy, 1960]. The questions '*How does this unity arise*?' and '*To what extent must it be considered a property of the organization of the organism, as opposed to a property emerging from its mode of life*?' remain open. A similar difficulty exists for the understanding of the functional organization of the nervous system, particularly if one considers the higher functions of man. Enumeration of the transfer functions of all nerve cells would leave us with a list, but not with a system capable of abstract thinking, description, and self-description. Such an approach would beg the question, '*How does the living organization give rise to cognition in general and to self-cognition in particular*?'

(b) Organisms are adapted to their environments, and it has appeared adequate to say of them that their organization represents the 'environment' in which they live, and that through evolution they have accumulated information about it, coded in their nervous systems. Similarly it has been said that the sense organs gather information about the 'environment', and through learning this information is coded in the nervous system [Cf. Young, 1967]. Yet this general view begs the questions, '*What does it mean to "gather information"*?' and '*What is coded in the genetic and nervous systems*?'.

A successful theory of cognition would answer both the epistemological and the biological questions. This I propose to do, and the purpose of this essay is to put forward a theory of cognition that should provide an epistemological insight into the phenomenon of cognition, and an adequate view of the functional organization of the cognizant organism that gives rise to such phenomena as conceptual thinking, language, and self-consciousness.

In what follows I shall not offer any formal definitions for the various terms used, such as 'cognition', 'life', or 'interaction', but I shall let their meaning appear through their usage. This I shall do because I am confident that the internal consistency of the theory will show that these terms indeed adequately refer to the phenomena I am trying to account for, and because I speak as an *observer*, and the validity of what I say at any moment has its foundation in the validity of the whole theory, which, I assert, explains why I can say it. Accordingly, I expect the complete work to give foundation to each of its parts, which thus appear justified only in the perspective of the whole.

Note: I shall be speaking of the organism as a unity, but when I wrote this essay I was not aware that the word unit did not always quite mean unity. Since I cannot now correct this. I beg the reader to bear this in mind.

II. THE PROBLEM

(1) Cognition is a biological phenomenon and can only be understood as such; any epistemological insight into the domain of knowledge requires this understanding.

(2) If such an insight is to be attained, two questions must be considered:

What is cognition as a function?

What is cognition as a process?

What follows should answer these two questions.

III. COGNITIVE FUNCTION IN GENERAL

THE OBSERVER

(1) Anything said is said by an observer. In his discourse the observer speaks to another observer, who could be himself; whatever applies to the one applies to the other as well. The observer is a human being, that is, a living system, and whatever applies to living systems applies also to him.

(2) The observer beholds simultaneously the entity that he considers (an organism, in our case) and the universe in which it lies (the organism's environment). This allows him to interact independently with both and to have interactions that are necessarily outside the domain of interactions of the observed entity.

(3) It is an attribute of the observer to be able to interact independently with the observed entity and with its relations; for him both are units of interaction (entities).

(4) For the observer an entity is an entity when he can describe it. To describe is to enumerate the actual or potential interactions and relations of the described entity. Accordingly, the observer can describe an entity only if there is at least one other entity from which he can distinguish it and with which he can observe it to interact or relate. This second entity that serves as a reference for the description can be any entity, but the ultimate reference for any description is the observer himself.

(5) The set of all interactions into which an entity can enter is its domain of interactions. The set of all relations (interactions through the observer) in which an entity can be observed is its domain of relations. This latter domain lies within the cognitive domain of the observer. An entity is an entity if it has a domain of interactions, and if this domain includes interactions with the observer who can specify for it a domain of relations. The observer can define an entity by specifying its domain of interactions; thus part of an entity, a group of entities, or their relations, can be made units of interactions (entities) by the observer.

(6) The observer can define himself as an entity by specifying his own domain of interactions; he can always remain an observer of these interactions, which he can treat as independent entities.

(7) *The observer is a living system and an understanding of cognition as a biological phenomenon must account for the observer and his role in it.*

THE LIVING SYSTEM

(1) Living systems are units of interactions; they exist in an ambience. From a purely biological point of view they cannot be understood independently of that part of the ambience with which they interact: the niche; nor can the niche be defined independently of the living system that specifies it.

(2) Living systems as they exist on earth today are characterized by exergonic metabolism, growth and internal molecular replication, all organized in a closed causal circular process that allows for evolutionary change in the way the circularity is maintained, but not for the loss of the circularity itself. Exergonic metabolism is required to provide energy for the endergonic synthesis of specific polymers (proteins, nucleic acids, lipids, polysaccharides) from the corresponding monomers, that is, for growth and replication; special replication procedures secure that the polymers synthesized be specific, that they should have the monomeric sequence proper to their class; specific polymers (enzymes) are required for the exergonic metabolism and the synthesis of specific polymers (proteins, nucleic acids, lipids, polysaccharides) [Cf. Commoner, 1965].

This circular organization constitutes a homeostatic system whose function is to produce and maintain this very same circular organization by determining that the *components* that specify it be those whose synthesis or maintenance it secures. Furthermore, this circular organization defines a living system as a unit of interactions and is essential for its maintenance as a unit; that which is not in it is external to it or does not exist. The circular organization in which the *components* that specify it are those whose synthesis or maintenance it secures in a manner such that the product of their functioning is the same functioning organization that produces them, is the living organization.

(3) It is the circularity of its organization that makes a living system a unit of interactions, and it is this circularity that it must maintain in order to remain a living system and to retain its identity through different interactions. All the peculiar aspects of the different kinds of organisms are superimposed on this basic circularity and are subservient to it, securing its continuance through successive interactions in an always changing environment. A living system defines through its organization the domain of all interactions into which it can possibly enter without losing its identity, and it maintains its identity only as long as the basic circularity that defines it as a unit of

interactions remains unbroken. Strictly, the identity of a unit of interactions that otherwise changes continuously is maintained only with respect to the observer, for whom its character as a unit of interactions remains unchanged.

(4) Due to the circular nature of its organization a living system has a self-referring domain of interactions (it is a self-referring system), and its condition of being a unit of interactions is maintained because its organization has functional significance only in relation to the maintenance of its circularity and defines its domain of interactions accordingly.

(5) Living systems as units of interactions specified by their condition of being living systems cannot enter into interactions that are not specified by their organization. The circularity of their organization continuously brings them back to the same internal state (same with respect to the cyclic process). Each internal state requires that certain conditions (interactions with the environment) be satisfied in order to proceed to the next state. Thus, the circular organization implies the prediction that an interaction that took place once will take place again. If this does not happen the system disintegrates; if the predicted interaction does take place, the system maintains its integrity (identity with respect to the observer) and enters into a new prediction. In a continuously changing environment these predictions can only be successful if the environement does not change in that which is predicted. Accordingly, the predictions implied in the organization of the living system are not predictions of particular events, but of classes of interactions. Every interaction is a particular interaction, but every prediction is a prediction of a class of interactions that is defined by those features of its elements that will allow the living system to retain its circular organization after the interaction, and thus, to interact again. This makes living systems inferential systems, and their domain of interactions a cognitive domain.

(6) The niche is defined by the classes of interactions into which an organism can enter. The environment is defined by the classes of interactions into which the observer can enter and which he treats as a context for his interactions with the observed organism. The observer beholds organism and environment simultaneously and he considers as the niche of the organism that part of the environment which he *observes* to lie in its domain of interactions. Accordingly, as for the observer the niche *appears* as part of the environment, for the observed organism the niche constitutes its entire domain of interactions, and as such it cannot be *part* of the environment that lies exclusively in the cognitive domain of the observer. Niche and environment, then, intersect only to the extent that the observer (including instruments) and the organism have comparable organizations, but even then

there are always parts of the environment that lie beyond any possibility of intersection with the domain of interactions of the organism, and there are parts of the niche that lie beyond any possibility of intersection with the domain of interactions of the observer. Thus for every living system its organization implies a prediction of a niche, and the niche thus predicted as a domain of classes of interactions constitutes its entire cognitive reality. If an organism interacts in a manner not prescribed by its organization, it does so as something different from the unit of interactions defined by its basic circularity, and this interaction remains outside its cognitive domain, although it may well lie within the cognitive domain of the observer.

(7) Every unit of interactions can participate in interactions relevant to other, more encompassing units of interactions. If in doing this a living system does not lose its identity, its niche may evolve to be contained by the larger unit of interactions and thus be subservient to it. If this larger unit of interactions is (or becomes) in turn also a self-referring system in which its components (themselves self-referring systems) are subservient to its maintenance as a unit of interactions, then it must itself be (or become) subservient to the maintenance of the circular organization of its components. Thus, a particular self-referring system may have the circular organization of a living system or partake functionally of the circular organization of its components, or both. The society of bees (the honey producing bees) is an example of a third order self-referring system of this kind; it has a circular organization superimposed on the second order self-referring systems that are the bees, which in turn have a circular organization superimposed on the first order living systems that are the cells; all three systems with their domains of interactions are subordinated both to the maintenance of themselves and to the maintenance of the others.

EVOLUTION

(1) Evolutionary change in living systems is the result of that aspect of their circular organization which secures the maintenance of their basic circularity, allowing in each reproductive step for changes in the way this circularity is maintained. Reproduction and evolution are not essential for the living organization, but they have been essential for the historical transformation of the cognitive domains of the living systems on earth.

(2) For a change to occur in the domain of interactions of a unit of interactions without its losing its identity with respect to the observer it must suffer an internal change. Conversely, if an internal change occurs in a unit of

interactions, without its losing its identity, its domain of interactions changes. A living system suffers an internal change without loss of identity if the predictions brought forth by the internal change are predictions which do not interfere with its fundamental circular organization. A system changes only if its domain of interactions changes.

(3) After reproduction the new unit of interactions has the same domain of interactions as the parental one only if it has the same organization. Conversely, the new unit of interactions has a different domain of interactions only if its organization is different, and hence, implies different predictions about the niche.

(4) Predictions about the niche are inferences about classes of interactions. Consequently, particular interactions which are indistinguishable for an organism may be different for an observer if he has a different cognitive domain and can describe them as different elements of a class defined by the conduct of the organism. The same applies to interactions that are identical for the organism but different for (have different effects) its different internal parts. Such interactions may result in different modifications of the internal states of the organism and, hence, determine different paths of change in its domain of interactions without loss of identity. These changes may bring about the production of offspring having domains of interactions different from the parental ones. If this is the case and a new system thus produced predicts a niche that cannot be actualized, it disintegrates; otherwise it maintains its identity and a new cycle begins.

(5) What changes from generation to generation in the evolution of living systems are those aspects of their organization which are subservient to the maintenance of their basic circularity but do not determine it, and which allow them to retain their identity through interactions; that is, what changes is the way in which the basic circularity is maintained, and not this basic circularity in itself. The manner in which a living system is compounded as a unit of interactions, whether by a single basic unit, or through the aggregation of numerous such units (themselves living systems) that together constitute a larger one (multicellular organisms), or still through the aggregation of these compound units that form self-referring systems of even higher order (insect societies, nations) is of no significance; what evolves is always a unit of interactions defined by the way in which it maintains its identity. The evolution of the living systems is the evolution of the niches of the units of interactions defined by their self-referring circular organization, hence, the evolution of the cognitive domains.

THE COGNITIVE PROCESS

(1) A cognitive system is a system whose organization defines a domain of interactions in which it can act with relevance to the maintenance of itself, and the process of cognition is the actual (inductive) acting or behaving in this domain. *Living systems are cognitive systems, and living as a process is a process of cognition.* This statement is valid for all organisms, with and without a nervous system.

(2) If a living system enters into a cognitive interaction, its internal state is changed in a manner relevant to its maintenance, and it enters into a new interaction without loss of its identity. In an organism without a nervous system (or its functional equivalent) its interactions are of a chemical or physical nature (a molecule is absorbed and an enzymatic process is initiated; a photon is captured and a step in photosynthesis is carried out). For such an organism the relations holding between the physical events remain outside its domain of interactions. The nervous system enlarges the domain of interactions of the organism by making its internal states also modifiable in a relevant manner by 'pure relations', not only by physical events; the observer sees that the sensors of an animal (say, a cat) are modified by light, and that the animal (the cat) is modified by a visible entity (say, a bird). The sensors change through physical interactions: the absorption of light quanta; the animal is modified through its interactions with the relations that hold between the activated sensors that absorbed the light quanta at the sensory surface. *The nervous system expands the cognitive domain of the living system by making possible interactions with 'pure relations'; it does not create cognition.*

(3) Although the nervous system expands the domain of interactions of the organism by bringing into this domain interactions with 'pure relations', the function of the nervous system is subservient to the necessary circularity of the living organization.

(4) The nervous system, by expanding the domain of interactions of the organism, has transformed the unit of interactions and has subjected acting and interacting in the domain of 'pure relations' to the process of evolution. As a consequence, there are organisms that include as a subset of their possible interactions, interactions with their own internal states (as states resulting from external and internal interactions) as if these were independent entities, generating the apparent paradox of including their cognitive domain within their cognitive domain. In us this paradox is resolved by what we call 'abstract thinking', another expansion of the cognitive domain.

(5) Furthermore, the expansion of the cognitive domain into the domain of 'pure relations' by means of a nervous system allows for non-physical interactions between organisms such that the interacting organisms orient each other toward interactions within their respective cognitive domains. Herein lies the basis for communication: the orienting behavior becomes a representation of the interactions toward which it orients, and a unit of interactions in its own terms. But this very process generates another apparent paradox: there are organisms that generate representations of their *own* interactions by specifying entities with which they interact as if these belonged to an independent domain, while as representations they only map their own interactions. In us this paradox is resolved simultaneously in two ways:

(a) We become *observers* through recursively generating representations of our interactions, and by interacting with several representations simultaneously we generate relations with the representations of which we can then interact and repeat this process recursively, thus remaining in a domain of interactions always larger than that of the representations.

(b) We become *self-conscious* through self-observation; by making descriptions of ourselves (representations), and by interacting with our descriptions we can describe ourselves describing ourselves, in an endless recursive process.

IV. COGNITIVE FUNCTION IN PARTICULAR

NERVE CELLS

(1) The neuron is the anatomical unit of the nervous system because it is a cell, and as such it is an independent integrated self-referring metabolic and genetic unit (a living system indeed).

(2) Anatomically and functionally a neuron is formed by a collector area (dendrites, and in some cases, also the cell body and part of the axon) united via a distributive element (the axon, and in some cases, also the cell body and main dendrites), capable of conducting propagated spikes to an effector area formed by the terminal branching of the axon. The functional state of the collector area depends on both its internal state (reference state) and on the state of activity of the effector areas synapsing on it. Correspondingly, the state of activity of the effector area depends on both the train of impulses generated at the corresponding collector area and on the pre-synaptic and non-synaptic interactions with distributive elements and other effector areas that may take place in the neuropil and in the immediate vicinity of the next collector areas. This is true even in the case of amacrine cells, in which the collector and effector areas may be intermingled. The distributive element determines where the effector exerts its influence.

(3) Whether one or two branches of a bifurcating axon are invaded by a nerve impulse propagating along it depends on their relative diameter and on the state of polarization of their membranes at their origin in the bifurcation zone. As a result, the pattern of effector activity, that is, the pattern of branch invasion which a train of impulses determines in the branches of the distributive element and effector area of a neuron, depends (i) on the spike interval distribution of the train of impulses, which determines the time that the axonal membrane at the branching zone has for recovery before the arrival of the next spike, and (ii) on the non-synaptic influences which, in the form of local water and ion movements caused by the electrical activity of neighboring elements, may produce diameter and polarization changes at the branching zones, and thus modify the invasibility of the branches by the arriving spikes.

(4) At any moment the state of activity of a nerve cell, as represented by

the pattern of impulses travelling along its distributive element, is a function of the spatio-temporal configuration of its input, as determined by the relative activity holding between the afferent neurons, that modulates the reference state proper of the collector area. It is known that in many neurons the recurrence of a given afferent spatio-temporal configuration results in the recurrence of the same state of activity, independently of the way in which such a spatio-temporal configuration is generated [Cf. Maturana and Frenk, 1963; Morrell, 1967]. [This is so in the understanding that two states of activity in a given cell are the 'same' (equivalent) if they belong to the same class, as defined by the pattern of impulses that they generate, and not because they are a one-to-one mapping of each other.] Also, the spatio-temporal configuration of the input to a neuron that causes in it the recurrence of a given state of activity is a class of afferent influences defined by a pattern in the relations holding between the active afferents and the collector; a given class of responses is elicited by a given class of afferent influences.

(5) For every nerve cell, at any moment, its transfer function at its collector area is a well-defined deterministic process [Cf. Segundo and Perkel, 1969]. Many neurons have several transfer functions, and different classes of afferent influences change their activity differently, causing them to generate different classes of activity in their effector areas. Because every nerve cell participates in the generation of the spatio-temporal configuration of afferent influences on the other nerve cells, all their states of activity must be considered as significant for their next states of activity. Thus there are two aspects to consider with respect to the activity of any given neuron: (i) its genesis, which must be considered in reference to the neuron itself and to the afferents to it; (ii) its participation in the generation of activity in other neurons for which it is an afferent influence, which must be considered in reference to those other neurous. In both cases the interactions between the neurons involved are strictly deterministic, although what is cause in one is not necessarily cause in another.

(6) The nerve impulses that travel along the distributive element originate at the point where this element emerges from the collector area. Each nerve impulse is the result of the state of excitation of the collector area at a given moment (as determined by the spatio-temporal configuration of the afferent excitatory and inhibitory influences acting upon it, and on its own internal generating mechanisms, if any) that spreads reaching a given threshold at the point of emergence of the distributor. Excitatory and inhibitory influences, however, do not superimpose linearly; their relative

participation in determining the production of nerve impulses, and hence, the state of activity of the neuron, depends on their relative spatial distribution on the collector area. Inhibition works by shunting off the spreading excitatory processes; as a result the relative contributions of a point of excitation and a point of inhibition in the generation of a nerve impulse depend on where, on the collector, they stand with respect to each other and with respect to the point of emergence of the distributive element. Excitation and inhibition must be seen as integral parts in the definition of the spatio-temporal configuration of afferent influences, not as independent processes. The shape of the collector area (its geometry) determines the class or classes of spatio-temporal configurations of afferent influences to which the cell responds.

(7) The neuropil is the site where the distributive elements and effector areas of many different neurons intermingle with each other and with the collector areas of the post-synaptic cells. Here non-synaptic interactions take place between neighboring elements which may cause in each other, as a result of the local movements of water and ions produced by their independent electrical activity, changes in diameter and polarization at their branching points. Depending on the time constant of these local changes, and on the capacity of the axons to homeostatically maintain their diameter at the new values, the pattern of branch invasion produced in a given effector area by a given train of impulses may be modified in a more or less permanent manner by these non-synaptic interactions. Something similar may happen during synaptic concomitances at the collector areas if synapses also affect each other non-synaptically, due to their spatial contiguity, causing each other more or less permanent changes in size (increase or decrease) and polarization (with the corresponding changes in effectiveness) as a result of their independent electrical activities. Thus, the neuropil may have to be seen as constituting a plastic system through which acquired self-addressing states of activity attain their functional significance as they become specified by the non-synaptic and synaptic concomitances generated by the interactions of the organism. It is not the repetition of the same state of activity which can cause neuronal changes of behavioral significance subordinated to the evolving domain of interactions of an organism, but rather it is the occurrence of local concomitant states of activity produced by seemingly unrelated interactions which can cause such subordinated changes in the reactive capacity of neurons.

(8) It follows that one should expect in a significant number of neurons, which may vary in different classes of animals according to the organization of their different neuropils, a continuous change in their transfer functions

(from collector to effector area), or in the circumstances under which they are activated, as a result of the past history of the organism. However, for the understanding of the functional organization of the nervous system it is necessary to consider that nerve cells respond at any moment with definite transfer functions to classes of afferent spatio-temporal configurations in their input, generating definite states of effector activity, and not to particular afferent states. Furthermore:

(a) Any interaction is represented in the nervous system by the sequence of states of relative neuronal activity leading to the conduct which it generates; this conduct should be repeatable to the extent that the interaction (sequence of states of relative activity) is reproducible, that is, as long as the historical transformation of the nervous system (learning) does not make it impossible.

(b) The nervous system always functions in the present, and it can only be understood as a system functioning in the present. The present is the time interval necessary for an interaction to take place; past, future and time exist only for the observer. Although many nerve cells may change continuously, their mode of operation and their past history can explain to the observer how their present mode of operation was reached, but not how it is realized now, or what their present participation in the determination of behavior is.

(c) Any behavior is defined through a sequence of states in the receptor surfaces (external and internal) that satisfy its direct or indirect subordination to the maintenance of the basic circularity of the living system. Since the nervous system is continuously changing through experience, what occurs when the observer sees a given behavior reenacted is a sequence of interactions that satisfy this subordination independently of the neuronal process which generated them. The more complex the domain of interactions of an organism, the more indirect is this subordination (an adequate mode of behavior subordinated to another), but not the less strict.

(d) An organism is a unit to the extent that its conduct results in the maintenance of its basic circularity (and hence identity), and two modes of conduct are equivalent if they satisfy the same class of requirements for this maintenance. For this reason an organism, as a self-regulated homeostatic organization, does not require a constant behavior in its deterministic component elements (in this case, neurons) if their changes become specified through the generation of conduct, and sameness of conduct is defined with respect to an observer or a function that must be satisfied.

Thus although at any moment every neuron functions deterministically with a definite transfer function, and generates a definite pattern of activity in its effector area, the transfer functions and the patterns of effector activity in many of them may change from one moment to another and the organism still will give rise to what the observer would call 'the same behavior'. The converse is also the case, and through what the observer would call 'different behaviors' the organism may satisfy its subordination to the same aspect of the maintenance of its basic circularity.

(9) From these notions it is apparent that the neuron cannot be considered as the functional unit of the nervous system; no neuron can have a fixed functional role in the generation of conduct if it must be continuously changing its participation in it. For the same reason a fixed collection of cells also cannot be considered as a functional unit of the nervous system. Only conduct itself can be considered as the functional unit of the nervous system.

(10) If nerve cells respond to classes of afferent configurations and not to particular afferent states, they must necessarily treat as equivalent particular afferent configurations that arise through interactions which for the observer are otherwise unrelated.

ARCHITECTURE

(1) In any given nervous system the great majority (and perhaps the totality) of its neurons can be assigned to well-defined morphological classes, each characterized by a given pattern of distribution of the collector and effector areas of its elements. As a result, the elements of the same class hold similar relations with each other and with other classes of neurons; the shapes of the nerve cells (collector area, distributive element, and effector area) specify their connectivity. These shapes are genetically determined and have been attained through evolution; the whole architecture of the brain is genetically determined and has been attained through evolution. The following implications are significant for the understanding of the nervous system:

(a) There is a necessary genetic variability in the shape of nerve cells as well as a variability that results from interactions of the organism with independent events during its development. The functional organization of the nervous system must be such as to tolerate this double variability.

(b) Due to the genetic and somatic variability no two nervous systems of animals of the same species (particularly if they have many cells) are identical, and they resemble each other only to the extent that they are organized according to the same general pattern. It is the organization defining the class, and not any particular connectivity, which determines the mode of functioning of any given kind of nervous system.

(2) The shapes of nerve cells and their packing are such that there is in general a great overlapping in the collector and effector areas of neurons of the same class. Also, the spatial distribution and the interconnections between different classes of neurons is such that any particular part of the nervous system is in general simultaneously related to many other parts; the parts interconnected, however, differ in different species, and as a result these have different interacting capabilities.

(3) The organism ends at the boundary that its self-referring organization defines in the maintenance of its identity. At this boundary there are sensors (the sensory surfaces) through which the organism interacts in the domain of relations and effectors (the effector surfaces) through which the nervous system modifies the posture of the organism in this domain. The sensory surfaces are in general constituted by collections of sensory elements (cells) with similar, though not identical, properties (classes of properties) which in their mode of interaction with the nervous system share the characteristics of neurons in general. As a result, whenever the organism enters into an interaction within the physical domain of interactions of the sensors, as a rule not one but many sensory elements are excited. The effectors are also multifarious and differ from each other in the manner in which they change the receptor surfaces of the organism during the interactions: action always leads to a change in the state of activity of the receptor surfaces.

(4) The architectural organization of the nervous system is subordinated to the order of the sensory and effector surfaces. This subordination has two aspects: (i) the receptor and effector surfaces project to the central nervous system retaining their proper topological relations; (ii) the topological relations specified by the receptor and effector surfaces in their projection constitute the basis for all the architectural order of the central nervous system. As a consequence, this architectural organization constitutes a system that interconnects these surfaces in a manner that permits the occurrence of certain concomitances of activity and not others in the different neuropils, and thus secures well-defined functional relations between these surfaces, specifying how they modify each other. *Truism*: the nervous system cannot give rise to a conduct that implies the concomitance of states of activity for which there is no anatomical basis. As a result of its architectural organization every point in the central nervous system constitutes an anatomical

localization with respect to the possibility of establishing certain functional concomitances. From this it follows that any localized lesion in the nervous system must necessarily interfere in a localized manner with the possibility of synthesizing some specific conduct (state of neural activity).

FUNCTION

(1) The way the nervous system functions is bound to its anatomical organization. The functioning of the nervous system has two aspects: one which refers to the domain of interactions defined by the nervous system (relations in general); the other which refers to the particular part of that domain used by a given species (particular classes of relations): Different species interact with different sets of relations (have different niches).

(2) The nervous system only interacts with relations. However, since the functioning of the nervous system is anatomy bound, these interactions are necessarily mediated by physical interactions; for an animal to discriminate objects visually the receptors in its eyes must absorb light quanta and be activated; yet, the objects that the animal sees are determined not by the quantity of light absorbed, but by the relations holding between the receptor-induced states of activity within the retina, in a manner determined by the connectivity of its various types of cells. Therefore, the nervous system defines through the relative weights of the patterns of interactions of its various components, both innate and acquired through experience, which relations will modify it at any given interaction [Cf. Maturana, 1965]. Or, in general, the organization and structure of a living system (its nervous system included) define in it a 'point of view', a bias or posture from the perspective of which it interacts, determining at any instant the possible relations accessible to its nervous system. Moreover, since the domain of interactions of the organism is defined by its structure, and since this structure implies a prediction of a niche, the relations with which the nervous system interacts are defined by this prediction and arise in the domain of interactions of the organism.

(3) Due to the properties of neurons, and due to the architecture of the nervous system, interactions within the nervous system give rise to activity in aggregates of cells. Also, for the same reasons, any given cell may assume the same state of activity under many different circumstances of interactions of the organism. Thus, under no circumstances is it possible to associate the activity of any particular cell with any particular interaction of the living

system. When any particular interaction takes place at the level of the sensors, the relations accessible to the nervous system are given at this level in a certain state of relative activity of the sensing elements and not in the state of activity of any particular one [Cf. Maturana, Uribe, and Frenk, 1968]. At the same time, although operational localizations can be established in the nervous system [Cf. Geschwind, 1965], these localizations are to be viewed in terms of areas where certain modalities of interactions converge, and not as localizations of faculties or functions. As a result of the mode of organization of the nervous system that I have emphasized, localized lesions should produce discrete functional deficiencies by impeding the convergence of activities necessary for the synthesis of a particular conduct (state of activity). The anatomical and functional organization of the nervous system secures the synthesis of behavior, not a representation of the world; hence, it is only with the synthesis of behavior that one can interfere. The nervous system is localized in terms of the organism's surfaces of interaction, but not in terms of representations of the interactions it can generate.

REPRESENTATION

(1) The fundamental anatomical and functional organization of the nervous system is basically uniform; the same functions and operations (excitation, inhibition, lateral interaction, recursive inhibition, etc.) are performed in its various parts, although in different contexts, and integrated in different manners. A partial destruction of the nervous system does not alter this basic uniformity, and, although the parts left untouched cannot do the same things that the whole did, they appear in their mode of operations identical to the untouched whole. To the observer, once the boundary of the sensors is passed, the nervous system, as a mode of organization, seems to begin at any arbitrary point that he may choose to consider; the answer to the question, '*What is the input to the nervous system*?' depends entirely on the chosen point of observation. This basic uniformity of organization can best be expressed by saying: all that is accessible to the nervous system at any point are states of relative activity holding between nerve cells, and all that to which any given state of relative activity can give rise are further states of relative activity in other nerve cells by *forming* those states of relative activity to which they respond. The effector neurons are not an exception to this since they, by causing an effector activity and generating an interaction, cause a change in the state of relative activity of the receptor elements at the receptor surfaces. This has a fundamental consequence: *unless they imply*

their origin (through concomitant events, their locations, or through the consequences of the new interactions which they originate) there is no possible distinction between internally and externally generated states of nervous activity.

(2) The relations with which the nervous system interacts are relations given by the physical interactions of the organism, and, hence, depend on its anatomical organization. For the observer the organism interacts with a given entity that he can describe in his cognitive domain. Yet, what modifies the nervous system of the observed organism are the changes in activity of the nerve cells associated with the sensing elements, changes that henceforth constitute an embodiment of the relations that arise through the interaction. These relations are not those that the observer can describe as holding between component properties of the entity in his cognitive domain; they are relations generated in the interaction itself and depend on both the structural organization of the organism and the properties of the universe that match the domain of interactions that this organization defines. Whenever such a relation recurs at the sensory surface, the same state of relative activity arises among the neurons in contact with the sensing elements. Two interactions that produce the same state of relative activity are identical for the nervous system, no matter how different they may be in the cognitive domain of the observer.

(3) Every relation is embodied in a state of relative activity of nerve cells, but also every state of relative activity acts to modify the relative activity of other nerve cells. Thus, relations through their embodiment in states of relative activity become units of internal interactions and generate additional relations, again embodied in states of relative activity which in turn may also become units of internal interactions, and so on, recursively.

(4) If an external interaction takes place, the state of activity of the nervous system is modified by the change in relative activity of the neurons, which in close association with the sensing elements embody the relations given in the interaction. Accordingly, that which the different states of activity thus generated can be said to represent are the relations given at the sensory surfaces by the interaction of the organism, and not an independent medium, least of all a description of an environment necessarily made in terms of entities that lie exclusively in the cognitive domain of the observer.

If an internal interaction takes place, the state of activity of the nervous system is modified by one of its own substates of relative activity that embodies one set of relations. However, that which the new state of relative activity represents is the relations given in the internal interaction, and not an independent set of relations or their description, in terms of some kind of entities, such as thoughts, that lie only within the cognitive domain of the observer.

(5) The classes of relations that can be embodied have been defined: (i) through the evolution of the general structural organization of the organism, and particularly, of the sensors, that has defined the classes of relation that are accessible to the nervous system; and (ii) through the evolution of a particular organization of the nervous system that defines for each class of animals (species) the specific mode of how these relations generate a behavior relevant to their maintenance.

(6) For any class of relations, the particular relations given as a result of a present interaction are embodied in a set of particular states of activity occurring in the present. This is the case independently of the history of the system. However, the relevance of the behavior generated by those states of activity for the maintenance of the living system is a function of history, and may depend both on the evolutionary history of the species and on the past experiences of the organism as an individual. In the first case I would speak of instinctive behavior, and in the second case of learned behavior. The description of learning in terms of past and present behavior lies in the cognitive domain of the observer; the organism always behaves in the present. The observer, however, by interacting with descriptions that he generates can treat interactions which do not recur as if they were in the present. This apparent paradox is resolved by generating the notion of time, past, present, and future, as a new expansion of the domain of interactions. Whenever an interaction takes place which is an element of a class experienced for the first time, it is sufficient that the state of activity which it generates be followed by the suppression of a peculiar concomitant internal state of activity (that is apparent in what the observer calls the emotion of anxiety or uncertainty) for the organism to experience the recurrence of an interaction of the same class, which takes place without such a concomitant state, as not new (in the sense that it can generate an established conduct as is apparent in the absence of anxiety) and, hence *known*. Any experience without anxiety can be described as known, and thus serve as a basis for the functional notion of time.

(7) There is no difference in the nature of the embodiment of the relations generated through either external or internal interactions; both are sets of states of neuronal activity that can be said to represent the interactions. In a nervous system capable of interacting with some of its own internal states as if they were independent entities, there are two consequences:

(a) The distinction between externally and internally generated interactions can only arise through a concomitance of events that indicates the source (sensory surface or not) of the state of activity caused by them, or through the outcome of new interactions which they initiate. A nervous system that is capable of treating its internally generated states of activity as different from its externally generated states, that is, of distinguishing their origin, is capable of abstract thinking.

(b) The nervous system can interact with the representations of its interactions (and hence, of the organism) in an endless recursive manner.

(8) Four comments:

(a) Notions such as embodiment of representation express the correspondence that the observer sees between relations, or sets of relations, and different states of activity of the nervous system, and, as such, lie in his cognitive domain. They describe the functional organization of the nervous system in the cognitive domain of the observer, and point to the ability of the nervous system to treat some of its own states as independent entities with which it can interact, but they do not characterize the nature of the functional subordination of the nervous system to its own states. This subordination is that of a functionally closed, state determined, ultrastable system, modulated by interactions [Cf. Ashby, 1960].

(b) The closed nature of the functional organization of the nervous system is a consequence of the self-referring domain of interactions of the living organization; every change of state of the organism must bring forth another change of state, and so on, recursively, always maintaining its basic circularity. Anatomically and functionally the nervous system is organized to maintain constant certain relations between the receptor and effector surfaces of the organism, which can only in that way retain its identity as it moves through its domain of interactions. Thus all conduct, as controlled through the nervous system, must (necessarily, due to the latter's architectural organization) lead through changes in the effector surfaces to specific changes in the receptor surfaces that in turn must generate changes in the effector surfaces that again . . . and so on, recursively. Conduct is thus a functional continuum that gives unity to the life of the organism through its transformations in the latter's self-referring domain of interactions. The evolutionary subordination of the architecture of the central nervous system to the topology of the sensory and effector surfaces appears as an obvious necessity.

(c) The ability of the nervous system to interact with its own internal states, as if these were independent entities, enters these internal states as modulating factors in the continuum of behavior. This requires an anatomical and functional internal reflection so that the internal organization of the nervous system can project itself onto itself retaining its morphological and functional topological relations, as the receptor and effector surfaces do in their own projection. This seems to have acquired an autonomous evolutionary course with the development of the neo-cortex in mammals, which arises as a center of internal anatomical projection, and whose evolution in this line is accompanied by an increased dependency of the organism on its own states of nervous activity.

(d) The closed nature of the functional organization of the nervous system (open only to modulations through interactions) is particularly evident in systematic observations that explicitly show the subordination of conduct to the correlation of activity between the receptor and effector surfaces [Cf. Held and Hein, 1963]. Experiments such as those of Held and Hein show that a cat does not learn to control its environment visually if raised in darkness and carried about only passively, by another cat, when under light. From these observations, it is apparent that the 'visual handling' of an environment is no handling of an environment, but the establishment of a set of correlations between effector (muscular) and receptor (proprioceptor and visual) surfaces, such that a particular state in the receptor surfaces may cause a particular state in the effector surfaces that brings forth a new state in the receptor surfaces . . . and so on. Behavior is like an instrumental flight in which the effectors (engines, flaps, etc.) vary their state to maintain constant, or to change, the readings of the sensing instruments according to a specified sequence of variations, which either is fixed (specified through evolution) or can be varied during the flight as a result of the state of the flight (learning). The same is apparent in the experiments with innate perception of depth [Cf. Gibson, 1950] that show that there is an innate system of correlations between certain states of the receptor and effector surfaces. The reference to a pre-established perception of depth is a description that lies in the cognitive domain of the observer, and as such only alludes to relations, through the observer, between elements that lie in his cognitive domain; but, as a process, this innate behavior obviously corresponds to one of optimization of sensory states.

DESCRIPTION

(1) A living system, due to its circular organization, is an inductive system

and functions always in a predictive manner: what happened once will occur again. Its organization, (genetic and otherwise) is conservative and repeats only that which works. For this same reason living systems are historical systems; the relevance of a given conduct or mode of behavior is always determined in the past. The goal state (in the language of the observer) that controls the development of an organism is, except for mutations, determined by the genome of the parent organism. The same is true for behavior in general; the present state is always specified from the previous state that restricts the field of possible modulations by independent concomitances. If a given state of relative activity in the nerve cells originates a given behavior, a recurrence of the 'same state' of relative activity should give rise to the 'same behavior' no matter how the recurrence originates. The relevance of such a behavior is determined by the significance that it has for the maintenance of the living organization, and it is in relation to this relevance that any subsequent behaviors are the same. With the expansion of the cognitive domain during evolution, the types of behavior have changed as well as how their relevance is implemented; different kinds of behavior are relevant to the maintenance of the basic circularity of the living organization through different domains of interactions, and hence, different fields of causal relations.

(2) Since the niche of an organism is the set of all classes of interactions into which it can enter, and the observer beholds the organism in an environment that he defines, for him any one of the organism's behaviors appears as an actualization of the niche, that is, as a first order description of the environment (henceforth denoted by a capital D: Description). This Description, however, is a description in terms of behavior (interactions) of the observed organism, not of representations of environmental states, and the relation between behavior and niche lies exclusively in the cognitive domain of the observer.

(3) An organism can modify the behavior of another organism in two basic ways:

(a) By interaction with it in a manner that directs both organisms toward each other in such a way that the ensuing behavior of each of them depends strictly on the following behavior of the other, e.g.: courtship and fight. A chain of interlocked behavior can thus be generated by the two organisms.

(b) By orienting the behavior of the other organism to some part of its domain of interactions different from the present interaction, but comparable to the orientation of that of the orienting organism. This can take place only if the domains of interactions of the two organisms are widely coincident; in this case no interlocked chain of behavior is elicited because the subsequent conduct of the two organisms depends on the outcome of independent, although parallel, interactions.

In the first case it can be said that the two organisms interact; in the second case that they communicate. The second case is the basis for any linguistic behavior; the first organism generates (as is apparent for the observer) a Description of its niche that, in addition to its own significance as a behavior (within the cognitive domain of the first organism, and independently of it), orients the second organism within its cognitive domain to an interaction from which ensues a conduct parallel to that of the first one, but unrelated to it. The conduct thus elicited by the orienting behavior is denotative: it points to a feature of the environment that the second organism encounters in its niche and Describes by the appropriate conduct, and that he can treat as an independent entity. The orienting behavior is, for the observer, a second order description (henceforth denoted by italics: *description*) that represents that which he considers it to denote. By contrast, the orienting behavior of the first organism is connotative for the second one, and implies for it an interaction within its cognitive domain which, if actualized, originates a behavior that Describes a particular aspect of its niche; that which an orienting behavior connotes is a function of the cognitive domain of the orientee, not the orienter.

(4) In an orienting interaction the behavior of the first organism, as a communicative *description* causes in the nervous system of the second one a specific state of activity; this state of activity embodies the relations generated in the interaction and represents the behavior of the second organism (Description of its niche) connoted by the orienting behavior of the first one. This representation, as a state of neuronal activity, can in principle be treated by the nervous system as a unit of interactions, and the second organism, if capable of doing so, can thus interact with representations of its own Descriptions of its niche as if these were independent entities. This generates yet another domain of interactions (and hence, another dimension in the cognitive domain), the domian of interactions with representations of behavior (interactions), orienting interactions included, as if these representations were independent entities within the niche: the *linguistic* domain.

(5) If an organism can generate a communicative *description* and then interact with its own state of activity that represents this *description*, generating another such *description* that orients towards this representation . . . , the process can in principle be carried on in a potentially infinite recursive

manner, and the organism becomes an *observer*: it generates discourse as a domain of interactions with representations of communciative *descriptions* (orienting behaviors).

Furthermore: if such an observer through orienting behavior can orient himself towards himself, and then generate communicative *descriptions* that orient him towards his *description* of this self-orientation, he can, by doing so recursively, *describe* himself *describing* himself . . . endlessly. Thus discourse through communicative *description* originates the apparent paradox of self-description: *self-consciousness*, a new domain of interactions.

(6) A nervous system capable of recursively interacting with its own states as if these were independent entities can do so regardless of how these states are generated, and in principle can repeat these recursive interactions endlessly. Its only limitation lies in the need that the progressive transformation of its actual and potential behavior, which in such a system is a necessary concomitant to behavior itself, be directly or indirectly subservient to the basic circularity of the living organization. The linguistic domain, the observer, and self-conciousness are each possible because they result as different domains of interactions of the nervous system with its own states in circumstances in which these states represent different modalities of interactions of the organism.

THINKING

(1) I consider that in a state-determined nervous system, the neurophysiological process that consists in its interacting with some of its own internal states as if these were independent entities corresponds to what we call thinking. Such internal states of nervous activity, otherwise similar to other states of nervous activity that participate in the specification of behavior, as in reflex mechanisms, cause conduct by determining specific changes of state in the nervous system. Thinking thus conceived, and reflex mechanisms, are both neurophysiological processes through which behavior emerges in a deterministic manner; they differ, however, in that in a reflex action we can, in our *description*, trace a chain of nervous interactions that begins with a specific state of activity at the sensory surfaces; while in thinking, the chain of nervous interactions that leads to a given conduct (change in the effector surfaces) begins with a distinguishable state of activity of the nervous system itself, whichever way it may have originated. Accordingly, thinking is a mode of operation of the nervous system that reflects functionally its internal anatomical projection (possibly multiply) onto itself.

(2) The process of thinking as characterized above is necessarily independent of language. That this is so even for what we call 'abstract thinking' in man is apparent from the observations of humans with split brains [Cf. Gazzaniga, Bogen and Sperry, 1965]. These observations show that the inability of the non-speaking hemisphere to speak does not preclude in it operations that the observer would call abstract thinking, and that the lack of language only implies that it cannot generate discourse. When we talk about concepts or ideas we *describe* our interactions with representations of our *descriptions*, and we think through our operation in the linguistic domain. The difficulty arises from our considering thinking through our *description* of it in terms or concepts as if it were something peculiar to man, and in some way isomorphic with the notions embodied in the *descriptions*, instead of attending to the functional process that makes these *descriptions* possible.

NATURAL LANGUAGE

(1) Linguistic behavior is orienting behavior; it orients the orientee within his cognitive domain to interactions that are independent of the nature of the orienting interactions themselves. To the extent that the part of its cognitive domain toward which the orientee is thus oriented is not genetically determined and becomes specified through interactions, one organism can in principle orient another to any part of its cognitive domain by means of arbitrary modes of conduct also specified through interactions. However, only if the domains of interactions of the two organisms are to some extent comparable, are such consensual orienting interactions possible and are the two organisms able to develop some conventional, but specific, system of communicative *descriptions* to orient each other to cooperative classes of interactions that are relevant for both.

(2) The understanding of the evolutionary origin of natural languages requires the recognition in them of a basic biological function which, properly selected, could originate them. So far this understanding has been impossible because language has been considered as a denotative symbolic system for the transmission of information. In fact, if such were the biological function of language, its evolutionary origin would demand the pre-existence of the function of denotation as necessary to develop the symbolic system for the transmission of information, but this function is the very one whose evolutionary origin should be explained. Conversely, if it is recognized that language is connotative and not denotative and that its function is to orient the orientee within his cognitive domain, and not to point to independent entities, it

becomes apparent that learned orienting interactions embody a function of non-linguistic origin that, under a selective pressure for recursive application, can originate through evolution the system of cooperative consensual interactions between organisms that is natural language. Particular orienting interactions, like any other learned conduct, arise from the substitution of one type of interaction for another as a cause for a given behavior, and their origin as a function of the general learning capacity of the nervous system is completely independent of the complexities of the system of cooperative interactions to which their recursive application gives rise. Widespread among animals other than man-orienting interactions are particularly evident in primates, in which it is easy to see how the audible and visible behavior of one individual orients others within their respective cognitive domains [Cf. Jay, 1968], and in dolphins which seem to have evolved a rich and efficient system of auditive cooperative interactions [Cf. Lilly, 1967]. In accordance with all this I maintain that learned orienting interactions, coupled with some mode of behavior that allowed for an independent recursive expansion of the domain of interactions of the organism, such as social life [Cf. Gardner and Gardner, 1969] and/or tool making and use, must have offered a selective basis for the evolution of the orienting behavior that in hominids led to our present-day languages.

(3) Behavior (function) depends on the anatomical organization (structure) of the living system, hence anatomy and conduct cannot legitimately be separated and the evolution of behavior is the evolution of anatomy and vice versa; anatomy provides the basis for behavior and hence for its variability; behavior provides the ground for the action of natural selection and hence for the historical anatomical transformations of the organism. Structure and function are, however, both relative to the perspective of interactions of the system and cannot be considered independently of the conditions that define it as a unit of interactions, for what is from one perspective a unit of interactions, from another may only be a component of a larger one, or may be several independent units. It is the dynamics of this process of individuation, as an historical process in which every state of a changing system can become a unit of interactions if the proper circumstances are given, what makes the evolution of living systems a deterministic process of necessarily increasing complication. Thus, in the evolution of language, natural selection, by acting upon orienting behavior as a function that if enhanced strongly increases the cooperation between social animals, has led to anatomical transformations which provide the basis for the increased complexity of the orienting conduct and the diversity of the interactions toward which man can be oriented in his

cognitive domain. The complexity of the orienting conduct has increased through an increase in the complexity and variety of motor behavior, particularly through vocalization and tool making. The diversity of the interactions toward which man can be oriented has increased through a concomitant expansion of the internal projection of the brain onto itself, by means of new interconnections between different cortical areas (as compared with other primates), between cortical areas and subcortical nuclei [Cf. Geschwind, 1964], and possibly also between different cortical layers and cellular systems within the cortex itself.

(4) So long as language is considered to be denotative it will be necessary to look at it as a means for the transmission of information, as if something were transmitted from organism to organism, in a manner such that the domain of uncertainties of the 'receiver' should be reduced according to the specifications of the 'sender'. However, when it is recognized that language is connotative and not denotative, and that its function is to orient the orientee within his cognitive domain without regard for the cognitive domain of the orienter, it becomes apparent that there is no transmission of information through language. It behooves the orientee, as a result of an independent internal operation upon his own state, to choose where to orient his cognitive domain; the choice is caused by the 'message', but the orientation thus produced is independent of what the 'message' represents for the orienter. In a strict sense then, there is no transfer of thought from the speaker to his interlocutor; the listener creates information by reducing his uncertainty through his interactions in his cognitive domain. Consensus arises only through cooperative interactions in which the resulting behavior of each organism becomes subservient to the maintenance of both. An observer beholding a communicative interaction between two organisms who have already developed a consensual linguistic domain, can describe the interaction as denotative; for him, a message (sign) appears as denotating the object which the conduct of the orientee *Describes* (specifies), and the conduct of the orientee appears determined by the message. However, because the outcome of the interaction is determined in the cognitive domain of the orientee regardless of the significance of the message in the cognitive domain of the orienter, the denotative function of the message lies only in the cognitive domain of the observer and not in the operative effectiveness of the communicative interaction. The cooperative conduct that may develop between the interacting organisms from these communicative interactions is a secondary process independent of their operative effectiveness. If it appears acceptable to talk about transmission of information in ordinary parlance, this is so

because the speaker tacitly assumes the listener to be identical with him and hence as having the same cognitive domain which he has (which never is the case), marvelling when a 'misunderstanding' arises. Such an approach is valid, for man created systems of communication where the identity of sender and receiver is implicitly or explicitly specified by the designer, and a message, unless disturbed during transmission, necessarily selects at the reception the same set of states that it represents at the emission, but not for natural languages.

(5) It behooves the interlocutor to choose where to orient in his cognitive domain as a result of a linguistic interaction. Since the mechanism of choice, as in every neuronal process, is state-dependent, the state of activity from which the choice (new state of neuronal activity) must arise restricts the possible choices and constitutes a reference background in the orientee. The same is valid for the speaker; the state of activity from which his communicative *description* (linguistic utterance) arises constitutes the reference background that specifies his choice. All the interactions that independently specify the reference background of each interlocutor constitute the context in which a given linguistic interaction takes place. Every linguistic interaction is thus necessarily context-dependent, and this dependency is strictly deterministic for both orienter and orientee, notwithstanding the different backgrounds of the two processes. It is only for the observer that there is any ambiguity in a linguistic interaction that he observes; this is because he has no access to the context in which it occurs. The sentence, 'They are flying planes,' is unambiguious for both interlocutors, regardless of the subsequent behavior which it originates in each of them; for the observer, however, who wants to predict the course of the ensuing interactions, it is ambiguous.

(6) If one considers linguistic interactions as orienting interactions it is apparent that it is not possible to separate, functionally, semantics and syntax, however separable they may seem in their description by the observer. This is true for two reasons:

(a) A sequence of communicative *descriptions* (words in our case) must be expected to cause in the orientee a sequence of successive orientations in his cognitive domain, each arising from the state left by the previous one. 'They are flying planes' clearly illustrates this; each successive word orients the listener to a particular interaction in his cognitive domain that is relevant in a particular manner (apparent in the conduct it generates) that depends on the previous orientation. The fact that it seems that the observer can more easily describe the word *are* (or any word) by referring to its grammatical and lexical functions, rather than by specifying the nature of the orientation that it causes (in terms of conduct or interactions), should not obscure the problem. The observer speaks, and any explanation of the word *are* that he may give lies in the descriptive domain, while the orientation caused by the word itself, as a change of state of the listener, is an internal interaction in *his* cognitive domain.

(b) An entire series of communicative *descriptions* can itself be a communicative *description*; the whole sequence once completed may orient the listener from the perspective of the state to which the sequence itself has led him. The limit to such complications lies exclusively in the capacity of the nervous system to discriminate between its own discriminable internal states, and to interact with them as if with independent entities.

(7) Linguistic behavior is an historical process of continuous orientation. As such, the new state in which the system finds itself after a linguistic interaction emerges from the linguistic behavior. The rules of syntax and generative grammar [Cf. Chomsky, 1968] refer to regularities that the observer sees in the linguistic behavior (as he *would* see in any behavior) which, arising from the functional organization of the system, specify the interactions that are possible at any given moment. Such rules, as rules, lie exclusively in the cognitive domain of the observer, in the realm of descriptions, because the transitions from state to state as internal processes in any system are unrelated to the nature of the interactions to which they give rise. Any correlation between different domains of interactions lies exclusively in the cognitive domain of the observer, as relations emerging from his simultaneous interactions with both.

(8) The coordinated states of neuronal activity which specify a conduct as a series of effector and receptor states whose significance arises in a consensual domain, does not differ in its neurophysiological generation from other coordinated states of neuronal activity which specify other conducts of innate or acquired significance (walking, flying, playing a musical instrument). Thus, however complex the motor and sensory coordinations of speech may be, the peculiarity of linguistic behavior does not lie in the complexity or nature of the series of effector and receptor states that constitute it, but in the relevance that such behavior acquires for the maintenance of the basic circularity of the interacting organisms through the development of the consensual domain of orienting interactions. Speaking, walking, or music-making do not differ in the nature of the coordinated neuronal processes which specify them but in the sub-domains of interactions in which they acquire their relevance.

(9) Orienting behavior in an organism with a nervous system capable of interacting recursively with its own states expands its cognitive domain by enabling it to interact recursively with *descriptions* of its interactions. As a result:

(a) Natural language has emerged as a new domain of interactions in which the organism is modified by its *descriptions* of its interactions, as they become embodied in states of activity of its nervous system, subjecting its evolution to its interactions in the domains of observation and self-consciousness.

(b) Natural language is necessarily generative because it results from the recursive application of the same operation (as a neurophysiological process) on the results of this application.

(c) New sequences of orienting interactions (new sentences) within the consensual domain are necessarily understandable by the interlocutor (orient him), because each one of their components has definite orienting functions as a member of the consensual domain that it contributes to define.

MEMORY AND LEARNING

(1) Learning as a process consists in the transformation through experience of the behavior of an organism in a manner that is directly or indirectly subservient to the maintenance of its basic circularity. Due to the state determined organization of the living system in general, and of the nervous system in particular, this transformation is an historical process such that each mode of behavior constitutes the basis over which a new behavior develops, either through changes in the possible states that may arise in it as a result of an interaction, or through changes in the transition rules from state to state. The organism is thus in a continuous process of becoming that is specified through an endless sequence of interactions with independent entities that select its changes of state but do not specify them.

(2) Learning occurs in a manner such that, for the observer, the learned behavior of the organism appears justified from the past, through the incorporation of a representation of the environment that acts, modifying its present behavior by recall; notwithstanding this, the system itself functions in the present, and for it learning occurs as an atemporal process of transformation. An organism cannot determine in advance when to change and when not to change during its flow of experience, nor can it determine in

advance which is the optimal functional state that it must reach; both the advantage of any particular behavior and the mode of behavior itself can only be determined *a posteriori*, as a result of the actual behaving of the organism subservient to the maintenance of its basic circularity.

(3) The learning nervous system is a deterministic system with a relativistic self-regulating organization that defines its domain of interactions in terms of the states of neuronal activity that it maintains constant, both internally and at its sensory surfaces, and that specifies these states at any moment through its functioning, and through the learning (historical transformation) itself. Consequently, it must be able to undergo a continuous transformation, both in the states it maintains constant, and in the way it attains them, so that every interaction in which new classes of concomitances occur effectively modifies it (learning curves) in one direction or the other. Since this transformation must occur as a continuous process of becoming without the previous specification of an end state, the final specification and optimization of a new behavior can only arise through the cumulative effect of many equally directed interactions, each of which selects, from the domain of structural changes possible to the nervous system in its structural dynamism, that which at that moment is congruent with its continued operation subservient to the basic circularity of the organism. Otherwise the organism disintegrates.

(4) The analysis of the nervous system made earlier indicated that the states of neuronal activity that arise in it through each interaction embody the relations given in the interaction, and not representations of the niche or the environment as the observer would describe them. This analysis also indicated that functionally such embodiments constitute changes in the reactivity of the nervous system, as a system closed on itself, to the modulating influences of further interactions. Consequently what the observer calls 'recall' and 'memory' cannot be a process through which the organism confronts each new experience with a stored representation of the niche before making a decision, but the expression of a modified system capable of synthesizing a new behavior relevant to its present state of activity.

(5) It is known that many neurons change their transfer functions as a result of the different concomitances of activity that occur in the neuropils of their collector and effector areas. Although it is not known what these changes are (development of new synapses or changes in their size, membrane changes, or changes in the pattern of spike invasion at the branching points of the axons), it can be expected from the relativistic organization of the nervous system that they should result in local morphological and functional changes that do not represent any particular interaction, but which permanently alter

the reactivity of the system. This anatomical and functional transformation of the nervous system must necessarily be occurring continuously as changes that the cells are able to stabilize with a permanency that lasts until the next modification, which can occur in any direction with respect to the previous one, or that subside by themselves after a certain number of interactions, but which are being locally triggered and selected through the actual concomitances of activity taking place in the neuropil itself.

(6) All changes in the nervous system during learning must occur without interference with its continued functioning as a self-regulating system; the unity that the observer sees in a living system throughout its continuous transformation is a strictly functional one. Accordingly, what appears constant for the observer when he ascertains that the same behavior is reenacted on a different occasion, is a set of relations that he defines as characterizing it, regardless of any change in the neurophysiological process through which it is attained, or any other unconsidered aspect of the conduct itself. Learning, as a relation between successive different modes of conduct of an organism such that the present conduct appears as a transformation of a past conduct arising from the recall of a specifiable past event, lies in the cognitive domain of the observer as a description of *his* ordered experiences. Likewise, memory as an allusion to a representation in the learning organism of its past experiences, is also a description by the observer of his ordered interactions with the observed organism; memory as a storage of representations of the environment to be used on different occasions in recall does not exist as a neurophysiological function.

(7) It is sufficient for a system to change its state after an interaction in a manner such that whenever a similar interaction recurs some internally determined concomitant state does not recur, although the same overt behavior is reenacted for it to treat two otherwise equivalent interactions as different elements of the same class. Such a peculiar state could be described as representing the emotional connotation of uncertainty which, present whenever a class of interactions is experienced for the first time, is suppressed after such an experience; the absence of such a concomitant state would suffice henceforth to treat differently (as known) all recurrent interactions of the same class. I maintain that modifications of this sort in the reactivity of the nervous system constitute the basis for the unidirectional ordering of experiences in a living system through 'recognition' without any storage of representations of the niche. First interactions that by error of the system are not accompanied by the above mentioned concomitant internal state (emotional connotation of uncertainty) would be treated as if known, as

occurs in the *déjà vu*. Conversely, interference with the suppression of the concomitant state of activity corresponding to this emotional connotation would result in the treatment of any recurrent interaction as if new (loss of recent memory).

(8) If such a system is capable of discourse, it will generate the temporal domain through the ascription of a unidirectional order to its experiences as they differ in their emotional connotations, and although it will continue to function in the present as an atemporal system, it will interact through its descriptions in the temporal domain. Past, present, and future, and time in general belong exclusively to the cognitive domain of the observer.

THE OBSERVER

Epistemological and Ontological Implications

(1) The cognitive domain is the entire domain of interactions of the organism. The cognitive domain can be enlarged if new modes of interactions are generated. Instruments enlarge our cognitive domain.

(2) The possibility of enlargement of the cognitive domain is unlimited; it is a historical process. Our brain, the brain of the observer, has specialized during evolution as an instrument for the discrimination of relations, both internally and externally generated relations, but relations given through and by interactions and embodied in the states of relative activity of its neurons. Furthermore, this occurs under circumstances in which the discriminations between states of relative activity – that for an observer represent the interactions of the organism, for the nervous system, that operate as a closed network – constitute only changes of relations of activity that arise between its components while it generates the internal and the sensory motor correlations that the states of the organism select. This has two aspects: one refers to the functional organization of the nerve cells which, with their responses, discriminate between different states of relative activity impinging upon them; the other refers to the ability of the nervous system, as a neuronal organization, to discriminate between its own states as these are distinguished and specified by the further states of activity that they generate. From this capacity of the nervous system to interact discriminately with its own states in a continuous process of self-transformation, regardless of how these states are generated, behavior emerges as a continuum of self-referred functional transformation. We cannot say in absolute terms what constitutes an input to our nervous system (the nervous system of the observer), because every

one of its states can be its input and can modify it as an interacting unit. We can say that every internal interaction changes us because it modifies our internal state, changing our posture or perspective (as a functional state) from which we enter into a new interaction. As a result new relations are necessarily created in each interaction and, embodied in new states of activity, we interact with them in a process that repeats itself as a historical and unlimited transformation.

(3) The observer generates a spoken *description* of his cognitive domain (which includes his interactions with and through instruments). Whatever *description* he makes, however, that *description* corresponds to a set of permitted states of relative activity in his nervous system embodying the relations given in his interactions. These permitted states of relative activity and those recursively generated by them are made possible by the anatomical and functional organization of the nervous system through its capacity to interact with its own states. The nervous system in turn has evolved as a system structurally and functionally subservient to the basic circularity of the living organization, and hence, embodies an inescapable logic: that logic which allows for a match between the organization of the living system and the interactions into which it can enter without losing its identity.

(4) The observer can *describe* a system that gives rise to a system that can *describe*, hence, to an observer. A spoken explanation is a paraphrase, a *description* of the synthesis of that which is to be explained; the observer explains the observer. A spoken explanation, however, lies in the domain of discourse. Only a full reproduction is a full explanation.

(5) The domain of discourse is a closed domain, and it is not possible to step outside of it through discourse. Because the domain of discourse is a closed domain it is possible to make the following ontological statement: *the logic of the* description *is the logic of the* describing *(living) system (and his cognitive domain)*.

(6) This logic demands a substratum for the occurrence of the discourse. We cannot talk about this substratum in absolute terms, however, because we would have to *describe* it, and a *description* is a set of interactions into which the *describer* and the listener can enter, and their discourse about these interactions will be another set of *descriptive* interactions that will remain in the same domain. Thus, although this substratum is required for epistemological reasons, nothing can be said about it other than what is meant in the ontological statement above.

(7) We as observers live in a domain of discourse interacting with *descriptions* of our *descriptions* in a recursive manner, and thus continuously generate

new elements of interaction. As living systems, however, we are closed systems modulated by interactions through which we define independent entities whose only reality lies in the interactions that specify them (their Description).

(8) For epistemological reasons we can say: there are properties which are manifold and remain constant through interactions. The invariance of properties through interactions provides a functional origin to entities or units of interactions; since entities are generated through the interactions that define them (properties), entities with different classes of properties generate independent domains of interactions: *no reductionism is possible.*

V. PROBLEMS IN THE NEUROPHYSIOLOGY OF COGNITION

(1) The observer can always remain in a domain of interactions encompassing his own interactions; he has a nervous system capable of interacting with its own states, which, by doing so in a functional context that defines these states as representations of the interactions from which they arise, allows him to interact recursively with representations of his interactions. This is possible because due to the general mode of organization of the nervous system there is no intrinsic difference between its internally and externally generated states of activity, and because each one of its specific states of activity is specifiable only in reference to other states of activity of the system itself.

(2) An organism with a nervous system capable of interacting with its own states is capable of *descriptions* and of being an observer if its states arise from learned orienting interactions in a consensual domain: it can *describe* its describing [Cf. Gardner and Gardner, 1969]. Through *describing* itself in a recursive manner, such an organism becomes a self-observing system that generates the domain of self-consciousness as a domain of self-observation. Self-consciousness then is not a neurophysiological phenomenon, it is a consensual phenomenon emerging in an independent domain of interactions from self-orienting behavior and lies entirely in the linguistic domain. The implications are twofold:

(a) The linguistic domain as a domain of orienting behavior requires at least two interacting organisms with comparable domains of interactions, so that a cooperative system of consensual interactions may be developed in which the emerging conduct of the two organisms is relevant for both. The specifiability through learning of the orienting interactions allows for a purely consensual (cultural) evolution in this domain, without it necessarily involving any further evolution of the nervous system; for this reason the linguistic domain in general, and the domain of self-consciousness in particular, are, in principle, independent of the biological substratum that generates them. However, in the actual becoming of the living system this independence is incomplete, on the one hand because the anatomical and neurophysiological organization of the brain, by determining the actual possibilities of confluence of different states of activity in it, specifies both the domain of possible interactions of the organism with relations and the complexity of the patterns of orienting interactions that it can distinguish, and on the other hand because the necessary subservience of the linguistic domain to the maintenance of the basic circularity of the organism through the generation of modes of behavior that directly or indirectly satisfy it limits the type of conduct that the organism can have without an immediate or eventual disintegration, or, of course, reduced rate of reproduction. Consequently, then, although the purely consensual aspects of the cultural evolution are independent of a simultaneous evolution of the nervous system, those aspects of the cultural evolution which depend on the possibility of establishing new classes of concomitances of activity in the nervous system, and generate new relations between otherwise independent domains, are not thus independent. Accordingly, once a cultural domain is established, the subsequent evolution of the nervous system is necessarily subordinated to it in the measure that it determines the functional validity of the new kinds of concomitances of activity that may arise in the nervous system through genetic variability.

(b) Since self-consciousness and the linguistic domain in general are not neurophysiological phenomena, it is impossible to account for them in terms of excitation, inhibition, networks, coding, or whatever else is the stuff of neurophysiology. In fact, the linguistic domain is fully explained only by showing how it emerges from the recursive application of orienting interactions on the results of their applications without being restricted as a domain by the neurophysiological substratum; what indeed is the problem is the need to account in purely physiological terms, without reference to meaning, for the synthesis of behavior in general, and for the synthesis of orienting behavior in particular. Accordingly, the fundamental quest in this respect should be to understand and explain

(i) *how does the nervous system interact with its own states, and is modified by them as if they were independent entities*?;

(ii) *how are these states specified neurophysiologically if they are defined by their own effectiveness in bringing forth certain internal or sensory states in the system*?;

(iii) *how is a given effector performance synthesized that is defined by the relative states of activity that it generates in the sensory surfaces and in the system itself*?; and

(iv) *how do the double or triple internal anatomical projections of the nervous system onto itself determine its capacity to single out some of its own states and interact with them independently?*.

(3) At any moment each nerve cell responds in a deterministic manner, and according to well defined transfer functions to classes of spatio-temporal activity caused at its collector area by the afferent influences impinging upon it; this occurs independently of how these afferent influences arise. This mode of cellular operation constitutes the basis for an associative process in which, whenever a given state of activity is produced in the nervous system, all neurons for which this state generates the proper classes of afferent influences enter into activity. Association thus conceived neurophysiologically is an inevitable process that calls into activity all cells that can be activated at any moment by a given state of the nervous system. No consideration of meaning enters into such a notion, since meaning, as a *description* by the observer, refers to the relevance that a mode of behavior has in the maintenance of the basic circularity of the organism as a consequence of self-regulation, and not in the mechanisms of the genesis of conduct. Association in terms of representations related by meaning lies in the cognitive domain of the observer exclusively. The nervous system is a system that functions maintaining constant certain states of relative activity, both internally and at the sensory surfaces, with reference only to some of its other states of relative activity. In this context the following considerations about its functional organization are significant:

(a) The nervous system can be described as a system that has evolved to specialize in the discrimination between states of neuronal relative activity (particularly in man) each of which is defined by the behavior it generates. This is valid for innate and learned behavior in circumstances in which every behavior is defined either by a set of states of activity maintained constant, or by their path of variation, both internally and at the sensory surfaces.

(b) The basic connectivity of the nervous system, and the original reactive capacity of the nerve cells, with which any animal is endowed by development, secures a basic pattern of flow for the nervous activity originating at any point in it. Thus, development specifies and determines both an initial repertoire of behavior over which all new conduct is built in a historical process of transformation, and an initial structurally specified set of possible associations that changes in an integrated manner with the historical transformation of behavior [Cf. Lorenz, 1966].

(c) Any modification of the transfer function of a nerve cell, resulting from new concomitances of activity, occurs modifying a preexisting behavior in a system that operates through maintaining invariant its definitory internal relations. In fact, any local change that would lead to the synthesis of a modified conduct by the organism, must be immediately accompanied by other changes arising through the adjustments that this must undergo in the process of maintaining constant its internal relations under its changed behavior. This is why it is the immediate relevance of a conduct for the maintenance of the organism in the present which at any moment selects the changes that take place during learning, and not the possible value of the conduct for future action.

(d) It is apparent that the nervous system cannot determine in advance the concomitances of activity under which it should change in a permanent manner; for it to satisfy future needs of the organism, it must operate under non-predictive changes continuously selected by the concomitances of activity arising in it. For this the nervous system must be capable of successful operation under the continuous transformation of its capacity to synthesize behavior, which necessarily results from a continuous change of the neurophysiological concomitances that determine the effective spatio-temporal configuration of activity impinging on the collector areas of its component neurons. Accordingly, it would seem of fundamental importance for the functional transformation of the system that many of its neurons should be able to change their relative participation in the synthesis of behavior as elements of different states of relative neuronal activity, independently of whether or not this is accompanied by any change in their transfer functions. In these circumstances the actual problem for the successful operation of the nervous system is the generation at any moment of the optimal configuration of activity necessary to synthesize a given behavior. However, since this continuous transformation of the functional capacity of the nervous system necessarily occurs under continuously successful behavior, such optimization requires no other specification than its attainment through the converging transformation of behavior itself.

(e) Since the nervous system is an inferential system, that is, since it functions as if any state that occurred once will occur again, a significant feature of its organization must be its necessary and continuous transformation as a function of the new concomitances of activity occurring in it. This functional requirement could be satisfied, for example, if any new local concomitance of activity in the neuropils changes the nerve cells in a deterministic and specific manner which does not represent any entity or event, but which modifies the neurophysiological circumstances under which the corresponding postsynaptic neurons are activated. Such can occur if the probability of spike invasion at the branching points of the afferent axons in the neuropils is permanently modified in one direction or another by the coincident novel activity in the neighboring structures, which, in the absence of synaptic interactions, cause, through local currents, local processes of growth or ungrowth in the branching zones of these axons. If this were the case four things would occur:

(i) The state of the nervous system would change, and hence, also its conduct, according to the new concomitances of activity produced in the neuropils through its different interactions.

(ii) Each state of activity of the system (as a state of relative neuronal activity) would be defined by the concomitances of activity in the neuropil that generate it, such that if they recur, it recurs.

(iii) Each new functional state of the neuropils would necessarily constitute the basis for their further modification, in such a manner that their morphological and functional organization would be under continuous historical transformation.

(iv) These changes in the neuropils would change the participation of the different neurons in the synthesis of behavior, independently of whether or not there are also changes in their transfer functions, by changing the circumstances of their activation. Accordingly, if an interaction (as described by the observer) recurs, no past conduct could be strictly reenacted by the organism, but this would have to synthesize a new adequate behavior that generates, in the context of its present interaction and in a manner that became specified through its structural transformation along its history of interactions, the internal and sensory motor correlations that maintain its identity.

(4) Learning is not a process of accumulation of representations of the environment; it is a continuous process of transformation of behavior through continuous change in the capacity of the nervous system to synthesize it. Recall does not depend on the indefinite retention of a structural invariant that represents an entity (an idea, image, or symbol), but on the functional ability of the system to create, when certain recurrent conditions are given, a behavior that satisfies the recurrent demands or that the observer would class as a reenacting of a previous one. As a consequence, the quest in the study of the learning process must answer two basic questions:

> '*What changes can a neuron undergo (in any of its component parts) which it can maintain constant for a certain time, and*

> *which modify in a definite manner its possible participation in different configurations of relative neuronal activity*?'; and
>
> '*What organization of the nervous system would permit continuous changes in the relative activity of its anatomical components, as a result of different concomitances in their activity, and still permit the synthesis of a conduct that is defined only by the states of relative neuronal activity that it generates, and not by the components used*?'.

(5) The nervous system is a strictly deterministic system whose structure specifies the possible modes of conduct that may emerge (be synthesized) from its functioning in a manner that varies according to the species, and the reactive perspective from which these modes of conduct may emerge. The reactive perspective, which the observer would call the emotional tone, does not specify a particular conduct, but determines the nature (aggressive, fearful, timid, etc.) of the course of the interaction [Cf. Kilmer, McCulloch and Blum, 1968]. Changes during development, maturation, hormonal action, drugs, or learning, do not modify the deterministic character of this organization but change the capacity that the system has at any moment to synthesize behavior. Furthermore, although any conduct or functional state always arises through a process of historical transformation from pre-existing modes of conduct or functional states, the nervous system functions in the present, and past history does not participate as an operant neurophysiological factor in the synthesis of conduct; nor does meaning, the relevance that a particular mode of conduct has, participate in it either. Time and meaning are effective factors in the linguistic domain, but as relational entities do not have neurophysiological correlates in the operation of the nervous system. Nor is the functional unity of the nervous system attained through a specific feature of its organization, but emerges from the functioning of its components (whatever these may be), each one to its own accord, under circumstances that define the ensemble as a unit of interactions in a particular domain [Cf. Lindauer, 1967, as an example in a social organism], and has no reality independent of these circumstances. Thus there is no peculiar neurophysiological process that could be shown to be responsible for this unity and to explain it. Furthermore, in a strict sense, although the nervous system has anatomical components it does not have functional parts since any mutilation leaves a functioning unit, with different properties as expressed by its possible interactions, but a unit in the corresponding domain. It appears incomplete only for the observer who beholds it as an entity from the perspective of

what he thinks it should be. Each component of the nervous system that the observer *describes* is defined in the domain of interactions of his observations, and as such is alien to the system which it is supposed to integrate. Every function has a structure which embodies it and makes it possible, but this structure is defined by the function in the domain of its operation as a set of relations between elements also defined in this domain. Neurons are the anatomical units of the nervous system, but are not the structural elements of its functioning. The structural elements of the functioning nervous system have not yet been defined, and it will probably be apparent when they are defined that they must be expressed in terms of invariants of relative activities between neurons, in some manner embodied in invariants of relations of interconnections, and not in terms of separate anatomical entities. In man-made systems this conceptual difficulty has not been so apparent because the system of relations (the theory) that integrates the parts that the describer (the observer) defines is provided by him, and is specified in his domain of interactions; as a consequence, these relations appear so obvious to the observer that he treats them as arising from the observation of the parts, and deludes himself, denying that he provides the unformulated theory that embodies the structure of the system which he projects onto them. In a self-referring system like a living system the situation is different: the observer can only make a *description* of his interactions with parts that he defines through interactions, but these parts lie in his cognitive domain only. Unless he explicitly or implicitly provides a theory that embodies the relational structure of the system, and conceptually supersedes his *description* of the components, he can never understand it. Accordingly, the full explanation of the organization of the nervous system (and of the organism) will not arise from any particular observation or detailed description and enumeration of its parts, but rather like any explanation, from the synthesis, conceptual or concrete, of a system that does what the nervous system (or the organism) does.

VI. CONCLUSIONS

The aim set forth in the introduction has been accomplished. Through the *description* of the self-referring circular organization of the living system, and through the analysis of the domains of interactions that such an organization specifies, I have shown the emergence of a self-referring system capable of making *descriptions* and of generating, through orienting interactions with other, similar, systems and with itself, both a consensual linguistic domain and a domain of self-consciousness, that is: I have shown the emergence of the observer. This result alone satisfies the fundamental demand put forth at the outset: '*The observer is a living system and any understanding of cognition as a biological phenomenon must account for the observer and his role in it*', and proves the validity of this analysis.

Although the answers to the various questions posed in the introduction and the fundamental implications of the analysis are to be found in the text itself to the extent that the theory adequately founds its whole development, there are several conclusions that I would like to state explicitly:

(i) The living organization is a circular organization which secures the production or maintenance of the *components* that specify it in such a manner that the product of their functioning is the very same organization that produces them. Accordingly, a living system is an homeostatic system whose homeostatic organization has its own organization as the variable that it maintains constant through the production and functioning of the *components* that specify it, and is defined as a unit of interactions by this very organization. It follows that living systems are a subclass of the class of circular and homeostatic systems. Also, it is apparent that the components referred to above cannot be specified as parts of the living system by the observer who can only subdivide a system in parts that he defines through his interactions, and which, necessarily, lie exclusively in his cognitive domain and are operationally determined by his mode of analysis. Furthermore, the relations through which the observer claims that these parts constitute a unitary system are relations that arise only through him by his simultaneous interactions with the parts and the intact system, and, hence, belong exclusively to his cognitive domain. Thus, although the observer can decompose a living system into parts that he defines, the *description* of these parts does not and cannot represent a living system. In principle a part should be definable through its relations within the unit that it contributes to form by its operation and interactions with other parts; this, however, cannot be attained because the analysis of a unit into parts by the observer destroys the very relations that would be significant for their characterization as effective components of the unit. Furthermore, these relations cannot be recovered through a *description* which lies in the cognitive domain of the observer and reflects only his interactions with the new units that he creates through his analysis. Accordingly, in a strict sense a unit does not have parts, and a unit is a unit only to the extent that it has a domain of interactions that defines it as different from that with respect to which it is a unit, and can be referred to only, as done above with the living system, by characterizing its organization through the domain of interactions which specify this distinction. In this context, the notion of *component* is necessary only for epistemological reasons in order to refer to the genesis of the organization of the unit through our *description*, but this use does not reflect the nature of its composition.

(ii) For every living system its particular case of self-referring circular organization specifies a closed domain of interactions that is its cognitive domain, and no interaction is possible for it which is not prescribed by this organization. Accordingly, for every living system the process of cognition consists in the creation of a field of behavior through its actual conduct in its closed domain of interactions, and not in the apprehension or the description of an independent universe. Our cognitive process (the cognitive process of the observer) differs from the cognitive processes of other organisms only in the kinds of interactions into which we can enter, such as linguistic interactions, and not in the nature of the cognitive process itself. In this strictly subject-dependent creative process, inductive inference is a necessary function (mode of conduct) that emerges as a result of the self-referring circular organization which treats every interaction and the internal state that it generates as if it were to be repeated, and as if an element of a class. Hence, functionally, for a living system every experience is the experience of a general case, and it is the particular case, not the general one, which requires many independent experiences in order that it be specified through the intersection of various classes of interactions. Consequently, although due to the historical transformation they have caused in organisms, or in their nervous systems, past interactions determine the inductive inferences that these make in the present, they do not participate in the inductive process itself. Inductive inference as a structural property of the living organization and of the thinking process, is independent of history, or of the relations between past and present that belong only to the domain of the observer.

(iii) Linguistic interactions orient the listener within his cognitive domain, but do not specify the course of his ensuing conduct. The basic function of language as a system of orienting behavior is not the transmission of information or the description of an independent universe about which we can talk, but the creation of a consensual domain of behavior between linguistically interacting systems through the development of a cooperative domain of interactions.

(iv) Through language we interact in a domain of *descriptions* within which we necessarily remain even when we make assertions about the universe or about our knowledge of it. This domain is both bounded and infinite; bounded because everything we say is a *description*, and infinite because every *description* constitutes in us the basis for new orienting interactions, and hence, for new *descriptions*. From this process of recursive application of *descriptions* self-consciousness emerges as a new phenomenon in a domain of self-description, with no other neurophysiological substratum than the neurophysiological substratum of orienting behavior itself. The domain of self-consciousness as a domain of recursive self-descriptions is thus also bounded and infinite.

(v) A living system is not a goal-directed system; it is, like the nervous system, a stable state-determined and strictly deterministic system closed on itself and modulated by interactions not specified through its conduct. These modulations, however, are apparent as modulations only for the observer who beholds the organism or the nervous system externally, from his own conceptual (*descriptive*) perspective, as lying in an environment and as elements in his domain of interactions. Contrariwise, for the functioning of the self-referring system itself all that there is is the sequence of its own self-subservient states. If this distinction is not made, one is liable to fail by including in the explanation of the organism and the nervous system features of interactions (*descriptions*) that belong exclusively to the cognitive domain of the observer.

(vi) It is tempting to talk about the nervous system as one would talk about a stable system with input. This I reject because it misses entirely the point by introducing the distortion of our participation as observers into the explanation of systems whose organization must be understood as entirely self-referring. What occurs in a living system is analogous to what occurs in an instrumental flight where the pilot does not have access to the outside world and must function only as a controller of the values shown in his flight instruments. His task is to secure a path of variations in the readings of his instruments, either according to a prescribed plan, or to one that becomes specified by these readings. When the pilot steps out of the plane he is bewildered by the congratulations of his friends on account of the perfect flight and landing that he performed in absolute darkness. He is perplexed because to his knowledge all that he did at any moment was to maintain the readings of his instruments within certain specified limits, a task which is in no way represented by the description that his friends (observers) make of his conduct.

In terms of their functional organization living systems do not have inputs and outputs, although under perturbations they maintain constant their set states, and it is only in our *descriptions*, when we include them as parts of larger systems which we define, that we can say that they do. When we adopt this *descriptive* approach in our analysis of the living organization we cannot but subordinate our understanding of it to notions valid only for man-made (allo-referring) systems, where indeed input and output functions are all important through the purposeful design of their role in the larger systems in which they are included, and this is misleading. In the organization of the living systems the role of the effector surfaces is only to maintain constant the set states of the receptor surfaces, not to act upon an environment, no matter how adequate such a description may seem to be for the analysis of adaptation, or other processes; a grasp of this is fundamental for the understanding of the organization of living systems.

(vii) The cognitive domain of the observer is bounded but unlimited; he can in an endless recursive manner interact with representations of his interactions and generate through himself relations between otherwise independent domains. These relations are *novelties* which, arising through the observer, have no other (and no less) effectiveness than that given to them by his behavior. Thus, he both creates (invents) relations and generates (specifies) the world (domain of interactions) in which he lives by continuously expanding his cognitive domain through recursive *descriptions* and representations of his interactions. The *new*, then, is a necessary result of the historical organization of the observer that makes of every attained state the starting point for the specification of the next one, which thus cannot be a strict repetition of any previous state; creativity is the cultural expression of this unavoidable feature.

(viii) The logic of the *description* and, hence, of *behavior* in general is, necessarily, the logic of the describing system; given *behavior as a referential and deterministic sequence of states of nervous activity in which each state determines the next one within the same frame of reference*, no contradiction can possibly arise in it as long as the latter remains unchanged by intercurrent interactions. If a change in the frame of reference takes place while a given *behavior* develops, a new one appears, such that the states following the change are determined with respect to it. If the new sequence of states (behavior) appears to an observer as contradicting the previous ones, this is so because he provides an independent and constant frame of reference in relation to which the successive sequences of states (behaviors) are contradictory. Such contradiction, however, lies exclusively in the cognitive domain of the observer, or of whatever provides the independent constant frame of reference. Contradictions (inconsistencies) then, do not arise in the generation of *behavior* but pertain to a domain in which the different *behaviors* acquire their significance by confronting an encompassing frame of reference through the interactions of the organism. Accordingly, thinking and discourse as modes of *behavior* are necessarily logically consistent in their generation, and that which the observer calls *rational* in them because they appear as concatenations of non-contradictory sequence dependent *descriptions*, is an expression of this necessary logical consistency. It follows that inconsistencies (*irrationalities*) in thinking and discourse as they appear to the observer arise from contextual changes in the circumstances that generate them while the independent frame of reference provided by the observer remains unchanged.

(ix) Due to the nature of the cognitive process and the function of the linguistic interactions, we cannot say anything about that which is independent of us and with which we cannot interact; to do that would imply a *description* and a *description* as a mode of conduct represents only relations given in interactions. Because the logic of the *description* is the same as the logic of the *describing* system we can assert the epistemological need for a substratum for the interactions to occur, but we cannot characterize this substratum in terms of properties independent of the observer. From this it follows that reality as a universe of independent entities about which we can talk is, necessarily, a fiction of the purely *descriptive* domain, and that we should in fact apply the notion of reality to this very domain of *descriptions* in which we, the *describing* system, interact with our descriptions as if with independent entities. This change in the notion of reality must be properly understood. We are used to talking about reality orienting each other through linguistic interactions to what we deem are sensory experiences of concrete entities, but which have turned out to be, as are thoughts and *descriptions*, states of relative activity between neurons that generate new *descriptions*. The question, '*What is the object of knowledge?*' becomes meaningless. There is no object of knowledge. To know is to be able to operate adequately in an individual or cooperative situation. We cannot speak about the substratum in which our cognitive behavior is given, and about that of which we cannot speak, we must remain silent, as indicated by Wittgenstein. This silence, however, does not mean that we fall into solipsism or any sort of metaphysical idealism. It means that we recognize that we, as thinking systems, live in a domain of *descriptions*, as has already been indicated by Berkeley, and that through *descriptions* we can indefinitely increase the complexity of our cognitive domain. Our view of the universe and of the questions we ask must change accordingly. Furthermore, this re-emergence of reality as a domain of *descriptions* does not contradict determinism and predictability in the different domains of interactions; on the contrary, it gives them foundation by showing that they are a necessary consequence of the isomorphism between the logic of the *description* and the logic of the *describing* system. It also shows that determinism and predictability are valid only within the field of this isomorphism; that is, they are valid only for the interactions that define a domain.

(x) The genetic and nervous systems are said to code information about the environment and to represent it in their functional organization. This is untenable; the genetic and nervous systems code processes that specify series of transformations from intial states, which can be decoded only through their actual implementation, not *descriptions* that the observer makes of an environment which lies exclusively in *his* cognitive domain [Cf. Bernal, 1965]. The following is an illustration of the problem:

Let us suppose that we want to build two houses. For such a purpose we hire two groups of thirteen workers each. We name one of the workers of the first group as the group leader and give him a book which contains all the plans of the house showing in a standard way the layout of walls, water pipes, electric connections, windows, etc., plus several views in perspective of the finished house. The workers study the plans and under the guidance of the leader construct the house, approximating continuously the final state prescribed by the *description*. In the second group we do not name a leader, we only arrange the workers in a starting line in the field and give each of them a book, the same book for all, containing only neighborhood instructions. These instructions do not contain words such as house, pipes, or windows, nor do they contain drawings or plans of the house to be constructed; they contain only instructions of what a worker should do in the different positions and in the different relations in which he finds himself as his position and relations change.

Although these books are all identical the workers read and apply different instructions because they start from different positions and follow different paths of change. The end result in both cases is the same, namely, a house. The workers of the first group, however, construct something whose final appearance they know all the time, while the workers of the second group have no views of what they are building, nor do they need to have obtained them even when they are finished. For the observer both groups are building a house, and he knows it from the start, but the house that the second group builds lies only in his cognitive domain; the house built by the first group, however, is also in the cognitive domains of the workers. The coding is obviously different in the two cases. In fact, the instructions contained in the book given to the first group clearly code the house as the observer would *describe* it, and the decoding task of the workers consists in purposefully doing things that will approximate to the construction of the *described* final state; this is why the house must be in their cognitive domain. In the second case, the instructions contained in each one of the thirteen identical books do not code a house. They code a process that constitutes a path of changing relationships which, if carried through under certain conditions, results in a system with a domain of interactions which has no intrinsic relationship with the beholding observer. That the observer should call this system a house is a feature of his cognitive domain, not of the system itself. In the first case the coding is isomorphic with a *description* of the house by the observer, and in fact constitutes a representation of it; in the second case it is not. The first case is typical of the way in which the observer codes the systems that he builds; the second corresponds to the way that the genome and nervous system constitute codes for the organism and for behavior, respectively, and one would never find in these codes any isomorphism with the *description* that the observer would make of the resultant systems with which he interacts. In what sense could one then say that the genetic and nervous systems code information about the environment? The notion of information refers to the observer's degree of uncertainty in his behavior within a domain of alternatives defined by him, hence the notion of information only applies within his cognitive domain. Accordingly, what one could at most say is that the genetic and nervous systems generate information through their self-specification when witnessed by the observer as if in their progressive self-decoding into growth and behavior.

(xi) There are different domains of interactions, and these different domains cannot explain each other because it is not possible to generate the phenomena of one domain with the elements of another; one remains in the same domain. One domain may generate the elements of another domain, but not its phenomenology, which in each domain is specified by the interactions of its elements, and the elements of a domain become defined only through the domain that they generate. Any nexus between different domains is provided by the observer who can interact as if with a single entity with the conjoined states of nervous activity generated in his brain by his concomitant interactions in several domains, or with independent *descriptions* of these interactions. Through these concomitant interactions in different domains (or with several *descriptions* within the *descriptive* domain) the observer generates relations between different domains (or between different *descriptions*) as states of neuronal activity that in him lead to definite modes of conduct (*descriptions*) that represent these conjoined interactions as singular independent entities. The number and kinds of relations the observer can generate in this manner is potentially infinite due to his recursive interactions with *descriptions*. Thus, relations, as states of neuronal activity arising from the concurrent interactions of the observer in different domains (physical and relational) constitute the elements of a new domain in which the observer interacts as a thinking system, but do not reduce one phenomenological domain into another. It is the simultaneous logical isomorphism of the new element (relations) with their source systems through their mode of origin (class intersection) that gives the new domain thus generated (*descriptions*) its explanatory capacity. An explanation is always a reproduction, either a concrete one through the synthesis of an equivalent physical system, or a conceptual one through a description from which emerges a system logically isomorphic to the original one, but never a reduction of one phenomenological domain into another. An adequate understanding of this irreducibility is essential for the comprehension of the biological phenomena, the consensual domains that living systems generate, and their conjoined evolution.

Many conclusions about self-consciousness and knowledge which arise from this mode of analysis have been proposed in one way or another by scientists and philosophers from their intuitive understanding, but never, to my knowledge, with an adequate biological and epistemological foundation. This I have done through the distinction between what pertains to the domain of the observer, and what pertains to the domain of the organism, and through carrying to their ultimate consequences the implications of the circular self-referring organization of the living systems: the implications of the functionally closed nature of the relativistic organization of the nervous system as a system under continuous transformation determined by relations of neuronal activity without the system ever stepping outside itself; and the implications of the non-informative orienting function of linguistic interactions. It is only after this has been done that the functional complexity of the living and linguistically interacting system can be properly grasped without its being concealed through such magic words as consciousness, symbolization, or information. Most of the detailed work is yet to be done, of course, but the fundamental first step of defining the perspective from which to look has here been taken. As a final remark, one could say what appears to be another paradox, but which points to the conceptual problem:

> *Living systems in general, and their nervous system in particular, are not made to handle a medium, although it has been through the evolution of their handling of their medium that they have become what they are, such that we can say what we can say about them.*

POST SCRIPTUM

No scientific work should be done without recognizing its ethical implications; in the present case the following deserve special attention:

(i) Man is a deterministic and relativistic self-referring autonomous system whose life acquires its peculiar dimension through self-consciousness; ethic and morality arise as commentaries that he makes on his behavior through self-observation. He lives in a continuously changing domain of descriptions that he generates through recursive interactions within that domain, and which has no other constant element in its historical transformation than his maintained identity as an interacting system. That is, man changes and lives in a changing frame of reference in a world continuously created and transformed by him. Successful interactions directly or indirectly subservient to the maintenance of his living organization constitute his only final source of reference for valid behavior within the domain of descriptions, and, hence, for truth; but, since living systems are self-referential systems, any final frame of reference is, necessarily, relative. Accordingly, no absolute system of values is possible and all truth and falsehood in the cultural domain are necessarily relative.

(ii) Language does not transmit information and its functional role is the creation of a cooperative domain of interactions between speakers through the development of a common frame of reference, although each speaker acts exclusively within his cognitive domain where all ultimate truth is contingent to personal experience. Since a frame of reference is defined by the classes of choices which it specifies, linguistic behavior cannot but be rational, that is, determined by relations of necessity within the frame of reference within which it develops. Consequently, no one can ever be rationally convinced of a truth which he did not have already implicitly in his ultimate body of beliefs.

(iii) Man is a rational animal that constructs his rational systems as all rational systems are constructed, that is, based on arbitrarily accepted truths (premises); being himself a relativistic self-referring deterministic system this cannot be otherwise. But if only a relative, arbitrarily chosen system of reference is possible, the unavoidable task of man as a self-conscious animal that can be an observer of its own cognitive processes is to explicitly choose a frame of reference for his system of values. This task he has always avoided by resorting to god as an absolute source of truth, or to self-delusion through reason, which can be used to justify anything by confusing the frames of reference and arguing in one domain with relations valid in another. The ultimate truth on which a man bases his rational conduct is necessarily subordinated to his personal experience and appears as an act of choice expressing a preference that cannot be transferred rationally; accordingly, the alternative to reason, as a source for a universal system of values, is aesthetic seduction in favor of a frame of reference specifically designed to comply with his desires (and not his needs) and defining the functions to be satisfied by the world (cultural and material) in which he wants to live.

AUTOPOIESIS

The Organization of the Living

1973

by

HUMBERTO R. MATURANA

and

FRANCISCO J. VARELA

TABLE OF CONTENTS

PREFACE

This small book is very large: it contains the living universe. It is a privilege to be asked to write this preface, and a delight to do so. That is because I recognize here a really important book, both in general and specifically. Before talking about the specific contents at all, I would like to explain why this is in general so.

IN GENERAL

We are the inheritors of categorized knowledge; therefore we inherit also a world view that consists of parts strung together, rather than of wholes regarded through different sets of filters. Historically, synthesis seems to have been too much for the human mind – where pratical affairs were concerned. The descent of the synthetic method from Plato through Augustine took men's perception into literature, art and mysticism. The modern world of science and technology is bred from Aristotle and Aquinas by analysis. The categorization that took hold of medieval scholasticism has really lasted it out. We may see with hindsight that the historic revolts against the scholastics did not shake free from the shackles of their reductionism.

The revolt of the rationalists – Descartes, Spinoza, Leibniz – began from a principle of 'methodical doubt'. But they became lost in mechanism, dualism, more and more categorization; and they ended in denying relation altogether. But relation is the stuff of system. Relation is the essence of synthesis. The revolt of the empiricists – Locke, Berkeley, Hume – began from the nature of understanding about the environment. But analysis was still the method, and categorization still the practical tool of advance. In the bizarre outcome, whereby it was the empiricists who denied the very existence of the empirical world, relation survived – but only through the concept of mental association between mental events. The system 'out there', which we call nature, had been annihilated in the process.

By the time Kant was devoting his prodigious mind to sorting all this out, the battle was lost. If the, quoting him, unconscious understanding organizes sensory experience into schemata, while conscious understanding organizes it into categories, the notion of identity remains for Kant forever transcendental.

Now the individual has vanished, in practical terms; as to the assemblage of individuals called society, that too has vanished into a transcendental construct. We have no need to legislate through any consensus of actual people, but only to meet needs that *might* have arisen from the noumenal will.

And what of science itself? Science is ordered knowledge. It began with classification. From Galen in the second century through to Linnaeus in the eighteenth, analysis and categorization provided the natural instrumentality of scientific progress. Ally this fact with the background of philosophical thought, and the scene is set for the inexorable development of the world view that is so difficult to challenge today. It is a world view in which real systems are annihilated in trying to understand them, in which relations are lost because they are not categorized, in which synthesis is relegated to poetry and mysticism, in which identity is a political inference. We may inspect the result in the structure and organization of the contemporary university.

It is an iron maiden, in whose secure embrace scholarship is trapped. For many, this is an entirely satisfactory situation, just because the embrace *is* secure. A man who can lay claim to knowledge about some categorized bit of the world, however tiny, which is greater than anyone else's knowledge of that bit, is safe for life: reputation grows, paranoia deepens. The number of papers increases exponentially, knowledge grows by infinitesimals, but understanding of the world actually recedes, because the world really is an interacting system. And since the world, in many of its aspects, is changing at an exponential rate, this kind of scholarship, rooted in the historical search of its own sanctified categories, is in large part unavailing to the needs of mankind.

There has been some recognition of this, and inter-disciplinary studies are by now commonplace in every university. But will this deal with the problem? Unfortunately, it will not. We still say that a graduate must have his 'basic discipline', and this he is solemnly taught – as if such a thing had a precise environmental correlate, and as if we know that God knew the difference between physics and chemistry. He learns also the academic *mores*, catches the institutional paranoia, and proceeds to propagate the whole business. Thus it is that an 'interdisciplinary study' often consists of a group of disciplinarians holding hands in a ring for mutual comfort. The ostensible topic has slipped down the hole in the middle. Among those who recognize this too, a natural enough debate has ensued on the subject: can an undergraduate be taught 'interdisciplinary studies' *as* his basic subject? But there is no such subject; there is no agreement on what it would be like; and we are rather short of anyone qualified to do the teaching. Those who resist the whole idea, in my view correctly, say that it would endanger the norms of good scholarship. There is a deadlock.

Against this background, let us consider *Autopoiesis*, and try to answer the question: 'What is it?' The authors say: "Our purpose is to understand the organization of living systems in relation to their unitary character". If the book deals with living systems, then it must be about biology. If it says anything scientific about organization, it must be about cybernetics. If it can recognize the nature of unitary character, it must be about epistemology – and also (remembering the first author's massive contribution to the understanding of perception) it will be about psychology too. Yes, it is indeed about all these things. Will you then call this an interdisciplinary study in the field of psychocyberbioepistemics? Do so only if you wish to insult the authors. Because *their* topic has *not* slipped down the hole in the middle. Therefore it is not an interdisciplinary study of the kind defined. It is not about analysis, but synthesis. It does not play the Game of the Categories. And it does not interrelate disciplines; it transcends them. If, because of my remarks about Kant, this seems to say that it annihilates them, then we are getting somewhere.

For there resides my belief in the book's general importance. The dissolution of the deadlock within the disciplinary system that I described above has got to be *meta*systemic, not merely interdisciplinary. We are not interested in forming a league of disciplinary paranoids, but (as Hegel could have told us) in a higher synthesis of disciplines. What emerges in this book is not classifiable under the old categories. Therefore it is predictable that no university could contain it, although all universities can and now do contain interdisciplinary institutions – because, in that very word, suitable obeisance is paid to the hallowed categories, and no one cares if the answers slip down the hole in the middle. As to the prediction that universities cannot contain this kind of work, I have often see it fulfilled. In the present case it is falsified, and I offer heartfelt congratulations to the University of Chile.

I say 'heartfelt' for this reason. In the mounting pile of new books printed every year that are properly called scientific, one may take hold of one's candle and search like a veritable Diogenes for a single one answering to the honest criteria I have proposed for a metasystemic utterance. There is only a handful in existence at all, which is not surprising in view of the way both knowledge and academia are organized. And yet, as I have also proposed, herein lies the world's real need. If we are to understand a newer and still evolving world; if we are to educate people to live in that world; if we are to legislate for that world; if we are to abandon categories and institutions that belong to a vanished world, as it is well-nigh desparate that we should; then knowledge must be rewritten. *Autopoiesis* belongs to the new library.

IN PARTICULAR

The authors first of all say that an autopoietic system is a homeostat. We already know what that is: a device for holding a critical systemic variable within physiological limits. They go on to the definitive point: in the case of autopoietic homeostasis, the critical variable is *the system's own organization*. It does not matter, it seems, whether every measurable property of that organizational structure changes utterly in the system's process of continuing adaptation. *It* survives.

This is a very exciting idea to me for two reasons. In the first place it solves the problem of identity which two thousand years of philosophy have succeeded only in further confounding. The search for the 'it' has led farther and farther away from anything that common sense could call reality. The 'it' of scholasticism is a mythological substance in which anything attested by the senses or testable by science inheres as a mere accident – its existence is a matter of faith. The 'it' of rationalism is unrealistically schizophrenic, because it is uncompromising in its duality – extended substance and thinking substance. The 'it' of empiricism is unrealistically insubstantial and ephemeral at the same time – *esse est percipi* is by no means the verdict of any experiencing human being.

The 'it' of Kant is the transcendental 'thing-in-itself' – an untestable inference, an intellectual gewgaw. As to the 'it' of science and technology in the twentieth century world of conspicuous consumption . . . 'it' seems to be no more than the collection of epiphenomena which mark 'it' as consumer or consumed. In this way hardheaded materialism seems to make 'it' as insubstantial as subjective idealism made it at the turn of the seventeenth century. And this, the very latest, the most down-to-earth, interpretation of 'it' the authors explicitly refute.

Their 'it' is notified precisely by its survival in a real world. You cannot find it by analysis, because its categories may all have changed since you last looked. There is no need to postulate a mystical something which ensures the preservation of identity *despite* appearances. The very continuation is 'it'. At least, that is my understanding of the authors' thesis – and I note with some glee that this means that Bishop Berkeley got the precisely right argument precisely wrong. He contended that something not being observed goes out of existence. Autopoiesis says that something that exists may turn out to be unrecognizable when you next observe it. This brings us back to reality, for that is surely true.

The second reason why the concept of autopoiesis excites me so much is that it involves the destruction of teleology. When this notion is fully worked out and debated, I suspect it will prove to be as important in the history of the philosophy of science as was David Hume's attack on causality. Hume considered that causation is a mental construct projected onto changing events which have, as we would say today, associated probabilities of mutual occurrence. I myself have for a long time been convinced that purpose is a mental construct imported by the observer to explain what is really an equilibrial phenomenon of polystable systems. The arguments in Chapter II appear to me to justify this view completely, and I leave the reader to engender his own excitement in the discovery of a 'purposelessness' that nonetheless makes good sense to a human being – just because he is allowed to keep his identity, which alone *is* his 'purpose'. It is enough.

But that salute to the authors is also self-congratulation, and I turn quickly aside. If a book is important, if at any time and from any source *information* is received, then something is changed – not merely confirmed. There are two arguments in this book that have changed me, and one of them effected its change after a profound inward struggle. Perhaps this part of the Preface should be printed as an epilogue: if I am not saying enough to be understood in advance of the reading, then I am sorry. It is too much to hope that the reader will return.

People who work with systems-theoretic concepts have often drawn attention to the subjective nature of 'the system'. A system is not something presented to the observer, it is something recognized by him. One of the consequences of this is that the labelling of connections between the system and its environment as either inputs or outputs is a process of arbitrary distinction. This is not very satisfatory. For example, a motor car in action is evidently a system. Suppose that it is recognized as 'a system for going from *A* to *B*'; then the water in the radiator is evidently an input, and displacement is evidently an output. Now consider the following scenario. Two men approach a motor car, and push it towards a second motor car. They then connect the batteries of the two cars with a pair of leads, and the engine of the first car fires. They disconnect the leads, and run the engine hard in neutral gear. We can guess what they are doing; but how is the objective scientist going to describe that system? Displacement is evidently an input, and one output is the rise in temperature of the water in the radiator. In case my example sounds too transparent, note that Aristotle thought that the brain was a 'human radiator', namely an apparatus for cooling the blood. Note also that he was right.

The fact is that we need a theoretical framework for any empirical investigation. This is the *raison d'être* of epistemology, and the authors make that point. In the trivial example I have just given, we need to know 'all about motor cars' before we can make sense of the empirical data. But it often happens in science that we know nothing at all about our 'motor cars', and sit there scratching our heads over data that relate to we know not what. There is a prime example of this in current scientific work, which is so embarrasssing that scientists in general pretend that it is not there. I am referring to the whole field of parapsychology – to the mass of data which seems to say: precognition, telepathy, telekinesis *exist*. But we flounder among statistical artifacts, and lack the theoretical framework for interpretation. This is made clear in the very name of ESP – 'extrasensory perception' which, if one thinks about it, constitutes an internal contradiction of terms.

Autopoiesis as a concept propounds a theoretical framework within which to cope with the confusion that arises from the subjective recognition of 'the system' and the arbitrary classification of its inputs and outputs. For the authors explain how we may treat autopoietic systems as if they were not autopoietic (that is, they are allopoietic) when the boundaries of the system are enlarged. Moreover, autopoietic systems may have allopoietic components. These ideas are immensely helpful, because our recognition of the circumstances in which a system should be regarded as either auto- or allo-poietic enables us to define 'the system' in an appropriate context. That is to say that the context is the recursion of systems within which the system we study is embedded, instead of being the cloud of statistical epiphenomena generated by our attempt to study it.

Understanding this changed me. The second change involved the intellectual struggle I mentioned earlier, and it concerns the authors' views on the information flowing within a viable system. In the numbered Paragraph (iv) of Section I of Chapter III they say: 'The notion of coding is a cognitive notion which represents the interactions of the observer, not a phenomenon operative in the physical domain. The same applies to the notion of regulation'. On first reading, this seemed to me plainly wrong. In the numbered Paragraph (v) of Section 3 of Chapter IV they say: "Notions such as coding, message or information are not applicable to the phenomenon of self-reproduction". Wrong again, I considered; indeed, outrageous – especially when taken with this remark from the first sentence of Section 3. 'reproduction . . . cannot enter as a defining feature of the organization of living systems'. Finally, in the numbered Paragraph (ii) of Section 3 of Chapter V, the authors say; 'A linguistic domain . . . is intrinsically non-informative'. Surely that is finally absurd?

All of this is totally alien to what we (most of us working in cybernetics) have believed. Information, including codes and messages and mappings, was indeed for us the whole story of the viable system. If one thinks of reproduction, for example, as the process of passing on a DNA code from an aging set of tissues to an embryonic set of tissues, then the survival of the code itself is what matters. The tissues of each generation are subject to aging and finally death, but the code is transmitted. The individual becomes insignificant, because the species is in the code. And that is why identity vanishes in an ageless computer program of bits – a program that specifies the hydrogen-bonded base pairs that link the sugar-phosphate backbones of the DNA molecule.

The whole outlook turns out to be wrong, and the book must speak for itself on this score. But it is an extraordinarily condensed book, which is why this preface is inordinately long. I do not know whether the authors' arguments about information led me to understand their concept of autopoiesis, or *vice versa*. What I am now sure about is that they are right. Nature is not about codes: we observers invent the codes in order to codify what nature is about. These discoveries are very profound.

What is less profound but equally important is the political consequence of this crisis about identity. The subordination of the individual to the species cannot be supported. "*Biology cannot be used any more to justify the dispensability of the individual for the benefit of the species, society or mankind, under the pretense that its role is to perpetuate them.*" After that, the world is a different place.

IN CONTENTION

The authors know it, and they draw the immediate inference. It is to say that scientists can no longer claim to be outside the social milieu within which they operate, invoking objectivity and disinterest; and in truth we have known this, or ought to have known it, ever since Hiroshima. But again this book gives us the theoretical basis for a view that might otherwise shroud something fundamental in a cloak of mere prudence. "No position or view that has any relevance in the domain of human relations can be deemed free from ethical and political implications, nor can a scientist consider himself alien to these implications". However, the authors go on to say that they do not fully agree between themselves on the questions this poses from the vantage point of their own work on autopoiesis – and they refuse to discuss them further (numbered Paragraph (iv) of Section 2 of Chapter V).

This seems to be because they do not resolve the question (posed a little earlier) whether human societies are or are not themselves biological systems. At this point, then, I ask to be relieved of the tasks of comment and interpretation; I ask for permission actively to enter this arena of discussion – where the angels fear to tread. For I am quite sure of the answer: yes, human societies *are* biological systems. Moreover, I claim that this book conclusively proves the point. This is a delicate matter, because presumably at least one of the originators of autopoietic theory disagrees, or is less than sure ... Nonetheless, I have read the book many times; and one of those readings was exclusively devoted to validating this contention against the authors' own criteria of autopoiesis at every point.

The outcome, to which I was admittedly predisposed because of my own work, says that any cohesive social institution is an autopoietic system – because it survives, because its method of survival answers the autopoietic criteria, and because it may well change its entire appearance and its apparent purpose in the process. As examples I list: firms and industries, schools and universities, clinics and hospitals, professional bodies, departments of state, and whole countries.

If this view is valid, it has extremely important consequences. In the first place it means that every social institution (in several of which any one individual is embedded at the intersect) is embedded in a larger social institution, and so on recursively – and that all of them are autopoietic. This immediately explains why the process of change at any level of recursion (from the individual to the state) is not only difficult to accomplish but actually impossible – in the full sense of the intention: 'I am going completely to change myself'. The reason is that the 'I', that self-contained autopoietic 'it', is a *component* of another autopoietic system. Now we already know that the first can be considered as allopoietic with respect to the second, and that is what makes the second a viable autopoietic system. But this is in turn means that the larger system perceives the embedded system as diminished – as less than fully autopoietic. That perception will be an illusion; but it does have consequences for the contained system. For now its own autopoiesis must respond to a special kind of constraint: treatment which attempts to deny its own autopoiesis.

Consider this argument at whatever level of recursion you please. An individual attempting to reform his own life within an autopoietic family cannot fully be his new self because the family insists that he is actually his old self. A country attempting to become a socialist state cannot fully become socialist; because there exists an international autopoietic capitalism in which it is embedded, by which the revolutionary country is deemed allopoietic. These conclusions derive from entailments of premises which the authors have placed in our hands. I think they are most valuable.

Then let me try to answer the obvious question: why do not the authors follow this line of development themselves, and write the second half of the book (as I hope they eventually will) – which would be about the nature and adaptation of social institutions, and the evolution of society itself? Well, to quote their sentence again: "Our purpose is to understand the organization of living systems in relation to their unitary character". This formulation of the problem begs the question as to what is allowed to be a called a living system, as they themselves admit. "Unless one knows which is the living organization, one cannot know which organization is living". They quickly reach the conclusion however (Subsection (b) of Section 2 of Chapter 1) that "*autopoiesis is necessary and sufficient to characterize the organization of living systems*". Then they display some unease, quoting the popular belief: ". . . and no synthetic system is accepted as living".

The fact is that if a social institution is autopoietic (and many seem to answer to the proper criteria) then, on the authors' own showing, it is necessarily alive. That certainly sounds odd, but it cannot be helped. It seems to me that the authors are holding at arms length their own tremendously important discovery. It does not matter about this mere word 'alive'; what does matter is that the social institution has *identity* in the biological sense; it is not just the random assemblage of interested parties that it is thought to be.

When it comes to social evolution then, when it comes to political change: we are not dealing with institutions and societies that will be different tomorrow because of the legislation we passed today. The legislation – even the revolution – with which we confront them does not alter them at all; it proposes a new challenge to their autopoietic adaptation. The behavior they exhibit may have to be very different if they are to survive: the point is that they have not lost their identities.

The interesting consequence is, however, that the way an autopoietic system will respond to a gross environmental challenge is highly predictable – *once the nature of its autopoiesis is understood*. Clever politicians intuit those adaptations; and they can be helped by good scientists using systems-theoretic models. Stupid politicians do not understand why social institutions do not lose their identities overnight when they are presented with perfectly logical reasons why they should; and these are helped by bad scientists who devote their effort to developing that irrelevant logic.

In an era when rapid institutional change is a prerequisite of peaceful survival in the face of every kind of exponentially rising threat, it seems to me that the architects of change are making the same mistake all over the world. It is that they perceive the system at their own level of recursion to be autopoietic, which is because they identify themselves with that system and know themselves to be so; but they insist on treating the systems their system contains, and those within which their system is contained, as allopoietic. This is allowable in terms of scientific description, when the input and output surfaces are correctly defined. Nonetheless it is politically blind to react towards the container and contained systems in a way which makes such a model evident, because at these other levels of recursion the relevant systems perceive themselves as autopoietic too.

This statement seems to be worth making. I could not have made it so succinctly without the language developed in this book. I could not have formulated it at all without the new concepts that Humberto Maturana and Francisco Varela have taught me. I thank them both very much, on behalf of everyone.

STAFFORD BEER

HUMBERTO R. MATURANA AND FRANCISCO J. VARELA

> So long as ideas of the nature of living things remain vague and ill-defined, it is clearly impossible, as a rule, to distinguish between an adaptation of the organism to the environment and a case of fitness of the environment for life, in the very most general sense. Evidently to answer such questions we must possess clear and precise ideas and definitions of living things. Life must by arbitrary process of logic be changed from the varying thing which it is into an independent variable or an invariant, shorn of many of its most interesting qualities to be sure, but no longer inviting fallacy through our inability to perceive clearly the questions involved.
>
> Henderson, *The Fitness of the Environment*

AUTOPOIESIS

The Organization of the Living

INTRODUCTION

A universe comes into being when a space is severed into two. A unity is defined. The description, invention and manipulation of unities is at the base of all scientific inquiry.

In our common experience we encounter living systems as unities that appear to us as autonomous entities of bewildering diversity endowed with the capacity to reproduce. In these encounters autonomy apears so obviously an essential feature of living systems that whenever something is observed that seems to have it, the naive approach is to deem it alive. Yet, autonomy, although continuously revealed in the self-asserting capacity of living systems to maintain their identity through the active compensation of deformations, seems so far to be the most elusive of their properties.

Autonomy and diversity, the maintenance of identity and the origin of variation in the mode in which this identity is maintained, are the basic challenges presented by the phenomenology of living systems to which men have for centuries addressed their curiosity about life.

In the search for an understanding of autonomy classic thought, dominated by Aristotle, created vitalism by endowing living systems with a non-material purposeful driving component that attained expression through the realization of their forms. After Aristotle, and as variations of his fundamental notions, the history of biology records many theories which attempt in one way or another to encompass all the phenomenology of living systems under some peculiar organizing force. However, the more biologists looked for the explicit formulation of one or other of these special organizing forces, the more they were disappointed by finding only what they could find anywhere else in the physical world: molecules, potentials and blind material interactions governed by aimless physical laws. Thence, under the pressure of unavoidable experience and the definite thrust of Cartesian thought a different outlook emerged, and mechanicism gradually gained the biological world by insisting that the only factors operating in the organization of living systems were physical factors, and that no non-matieral vital organizing force was necessary. In fact, it seems now apparent that any biological phenomenon, once properly defined, can be described as arising from the interplay of physicochemical processes whose relations are specified by the context of its definition.

Diversity has been removed as a source of bewilderment in the understanding of the phenomenology of living systems by Darwinian thought and particulate genetics which have succeeded in providing an explanation for it and its origin without resorting to any peculiar directing force. Yet, the influence of these notions through their explanation of evolutionary change, has gone beyond the mere accounting for diversity: it has shifted completely the emphasis in the evaluation of the biological phenomenology from the individual to the species, from the unity to the origin of its parts, from the present organization of living systems to their ancestral determination.

Today the two streams of thought represented by the physicochemical and the evolutionary explanations, are braided together. The molecular analysis seems to allow for the understanding of reproduction and variation, the evolutionary analysis seems to account for how these processes might have come into being. Apparently we are at a point in the history of biology where the basic difficulties have been removed. Biologists, however, are uncomfortable when they look at the phenomenology of living systems as a whole. Many manifest this discomfort by refusing to say what a living system is. Others attempt to encompass present ideas under comprehensive theories governed by organizing notions, like cybernetic principles, that require from the biologists the very understanding that they want to provide. The ever present question is: 'What is common to all living systems that we qualify them as living'; if not a vital force, if not an organizing principle of some kind, what then? To take only a notable recent example let us mention J. Monod's book *Le hasard et la necessitè*. He tries to answer this question but, following the emphasis of evolutionary thought, he postulates a teleonomic organization of molecular nature and the subordination of the organization of the individual to a plan defined by the species, in which the invariance of reproduction is determinant. Yet, teleonomic and evolutionary notions leave the question of the nature of the organization of the living unity essentially untouched.

Our endeavor is to disclose the nature of the living organization. However, in our approach we make a starting point of the unitary character of a living system, and maintain that the evolutionary thought through its emphasis on diversity, reproduction and the species in order to explain the dynamics of change has obscured the necessity of looking at the autonomous nature of living unities for the understanding of the biological phenomenology. Also we think that the maintenance of identity and the invariance of defining relations in the living unities are at the base of all possible ontogenic and evolutionary transformation in biological systems, and this we intend to explore. Thus, our purpose is: to understand the organization of living systems in relation to their unitary character.

Our approach will be mechanistic: no forces or principles will be adduced which are not found in the physical universe. Yet, our problem is the living organization and therefore our interest will not be in properties of components, but in processes and relations between processes realized through components. This is to be clearly understood. An explanation is always a reformulation of a phenomenon showing how its components generate it through their interactions and relations. Furthermore, an explanation is always given by us as observers, and it is central to distinguish in it what pertains to the system as constitutive of its phenomenology from what pertains to our domain of description, and hence to our interactions with it, its components and the context in which it is observed. Since our descriptive domain arises because we simultaneously behold the unity and its interactions in the domain of observation, notions arising in the domain of description do not pertain to the constitutive organization of the unity (phenomenon) to be explained. Furthermore, an explanation may take different forms according to the nature of the phenomenon explained. Thus, to explain the movement of a falling body one resorts to properties of matter, and to laws that describe the conduct of material bodies according to these properties (kinetic and gravitational laws), while to explain the organization of a control plant one resorts to relations and laws that describe the conduct of relations. In the first case, the elements used in the explanation are bodies and their properties; in the second case, they are relations and their relations, independently of the nature of the bodies that satisfy them. As in this latter case, in our explanation of the organization of living systems, we shall be dealing with the relations which the actual physical components must satisfy to constitute one, not with the identification of these components. It is our assumption that there is an organization that is common to all living systems, whichever the nature of their components. Since our subject is this organization, not the particular ways in which it may be realized, we shall not make distinctions between classes or types of living systems.

This mode of thinking is not new, and is explicitly related to the very name of mechanicism. We maintain that living systems are machines and by doing this we point at several notions which should be made explicit. First, we imply a non-animistic view which it should be unnecessary to discuss any further. Second, we are emphasizing that a living system is defined by its organization and, hence, that it can be explained as any organization is explained, that is, in terms of relations, not of component properties. Finally, we are pointing out from the start the dynamism apparent in living systems and which the word 'machine' connotes.

We are asking, then, a fundamental question: 'What is the organization of living systems, what kind of machines are they, and how is their phenomenology, including reproduction and evolution, determined by their unitary organization?'

CHAPTER I

ON MACHINES, LIVING AND OTHERWISE

1. MACHINES

Machines are usually viewed as concrete hardware systems, defined by the nature of their components and by the purpose that they fulfill in their operations as man-made artifacts. This view however is obviously naive because it says nothing about how they are constituted. That machines are unities is apparent; that they are made of components that are characterized by certain properties capable of satisfying certain relations that determine in the unity the interactions and transformations of these same components is also apparent. What is not so apparent is that the actual nature of the components, and the particular properties that these may possess other than those participating in the interactions and transformations which constitute the system, are irrelevant and can be any. In fact, the significant properties of the components must be taken in terms of relations, as the network of interactions and transformations into which they can enter in the working of the machine which they integrate and constitute as a unity.

The relations that define a machine as a unity, and determine the dynamics of interactions and transformations which it may undergo as such a unity, constitute the *organization* of the machine. The actual relations which hold among the components which integrate a concrete machine in a given space, constitute its *structure*. The organization of a machine (or system) does not specify the properties of the components which realize the machine as a concrete system, it only specifies the relations which these must generate to constitute the machine or system as a unity. Therefore, the organization of a machine is independent of the properties of its components which can be any, and a given machine can be realized in many different manners by many different kinds of components. In other words, although a given machine can be realized by many different structures, for it to constitute a concrete entity in a given space its actual components must be defined in that space, and have the properties which allow them to generate the relations which define it.

The use to which a machine can be put by man is not a feature of the organization of the machine, but of the domain in which the machine operates, and belongs to our description of the machine in a context wider than the machine itself. This is a significant notion. Man made machines are all made with some purpose, practical or not, but with some aim (even if it is only to amuse) that man specifies. This aim usually appears expressed in the product of the operation of the machine, but not necessarily so. However, we use the notion of purpose when talking of machines because it calls into play the imagination of the listener and reduces the explanatory task in the effort of conveying to him the organization of a particular machine. In other words, with the notion of purpose we induce the listener to invent the machine we are talking about. This, however, should not lead us to believe that purpose, or aim, or function, are constitutive properties of the machine which we describe with them; such notions are intrinsic to the domain of observation, and cannot be used to characterize any particular type of machine organization. The product of the operations of a machine, however, can be used to this end in a non-trivial manner in the domain of descriptions generated by the observer.

2. LIVING MACHINES

That living systems are machines cannot be shown by pointing to their components. Rather, one must show their organization in a manner such that the way in which all their peculiar properties arise, becomes obvious. In order to do this, we shall first characterize the kind of machines that living systems are, and then show how the peculiar properties of living systems may arise as consequences of the organization of this kind of machines.

a. *Autopoietic machines*

There are machines which maintain constant, or within a limited range of values, some of their variables. The way this is expressed in the organization of these machines must be such as to define the process as occurring completely within the boundaries of the machine which the very same organization specifies. Such machines are homeostatic machines and all feedback is internal to them. If one says that there is a machine *M*, in which there is a feedback loop through the environment so that the effects of its output affect its input, one is in fact talking about a larger machine *M'* which includes the environment and the feedback loop in its defining organization.

Autopoietic machines are homeostatic machines. Their peculiarity, however, does not lie in this but in the fundamental variable which they maintain constant. *An autopoietic machine is a machine organized* (*defined as a unity*) *as a network of processes of production (transformation and destruction) of components that produces the components which:* (i) *through their interactions and transformations continuously regenerate and realize the network of processes (relations) that produced them; and* (ii) *constitute it (the machine) as a concrete unity in the space in which they (the components) exist by specifying the topological domain of its realization as such a network.* It follows that an autopoietic machine continuously generates and specifies its own organization through its operation as a system of production of its own components, and does this in an endless turnover of components under conditions of continuous perturbations and compensation of perturbations. Therefore, an autopoietic machine is an homeostatic (or rather a relations-static) system which has its own organization (defining network of relations) as the fundamental variable which it maintains constant. This is to be clearly understood. Every unity has an organization specifiable in terms of static or dynamic relations between elements, processes, or both. Among these possible cases, autopoietic machines are unities whose organization is defined by a particular network of processes (relations) of production of components, the autopoietic network, not by the components themselves or their static relations. Since the relations of production of components are given only as processes, if the processes stop, the relations of production vanish; as a result, for a machine to be autopoietic, its defining relations of production must be continuously regenerated by the components which they produce. Furthermore, the network of processes which constitute an autopoietic machine is a unitary system in the space of the components that it produces and which generate the network through their interactions. The autopoietic network of processes, then, differentiates autopoietic machines from any other kind of unit. In fact: (i) in a man-made machine in the physical space, say a car, there is an organization given in terms of a concatenation of processes, yet, these processes are not processes of production of the components which specify the car as a unity since the components of a car are produced by other processes which are independent of the organization of the car and its operation. Machines of this kind are non-autopoietic dynamic systems. (ii) In a natural physical unity like a crystal, the spatial relations among the components specify a lattice organization which defines it as a member of a class (a crystal of a particular kind), while the kinds of components which constitute it specify it as a particular case in that class. Thus, the organization of a crystal is specified by the spatial relations which define the relative position of its components, while these specify its unity in the space in which they exist – the physical space. This is not so with an autopoietic machine. In fact, although we find spatial relations among its components whenever we actually or conceptually freeze it for an observation, the observed spatial relations do not (and cannot) define it as autopoietic. This is so because the spatial relations between the components of an autopoietic machine are specified by the network of processes of production of components which constitute its organization and they are therefore necessarily in continuous change. A crystal organization then, lies in a different domain than the autopoietic organization: a domain of relations between components, not of relations between processes of production of components; a domain of processes, not of concatenation of processes. We normally acknowledge this by saying that crystals are static.

It is important to realize that we are not using the term organization in the definition of an autopoietic machine in a mystical or transcendental sense, pretending that it has any explanatory value of its own. We are using it only to refer to the specific relations that define an autopoietic system. Thus, autopoietic organization simply means processes interlaced in the specific form of a network of productions of components which realizing the network that produced them constitute it as a unity. It is for this reason that we can say that every time that this organization is actually realized as a concrete system in a given space, the domain of the deformations which this system can withstand without loss of identity while maintaining constant its organization, is the domain of changes in which it exists as a unity. It is thus clear that the fact that autopoietic systems are homeostatic systems which have their own organization as the variable that they maintain constant, is a necessary consequence of the autopoietic organization.

The consequences of this autopoietic organization are paramount:

(i) Autopoietic machines are autonomous; that is, they subordinate all changes to the maintenance of their own organization, independently of how profoundly they may otherwise be transformed in the process. Other machines, henceforth called allopoietic machines, have as the product of their functioning something different from themselves (as in the car example). Since the changes that allopoietic machines may suffer without losing their definitory organization are necessarily subordinated to the production of something different from themselves, they are not autonomous.

(ii) Autopoietic machines have individuality; that is, by keeping their organization as an invariant through its continuous production they actively maintain an identity which is independent of their interactions with an observer. Allopoietic machines have an identity that depends on the observer and is not determined through their operation, because its product is different from themselves; allopoietic machines do not have individuality.

(iii) Autopoietic machines are unities because, and only because, of their specific autopoietic organization: their operations specify their own boundaries in the processes of self-production. This is not the case with an allopoietic machine whose boundaries are defined by the observer, who by specifying its input and output surfaces, specifies what pertains to it in its operations.

(iv) Autopoietic machines do not have inputs or outputs. They can be perturbated by independent events and undergo internal structural changes which compensate these perturbations. If the perturbations are repeated, the machine may undergo repeated series of internal changes which may or may not be identical. Whichever series of internal changes takes place, however, they are always subordinated to the maintenance of the machine organization, condition which is definitory of the autopoietic machines. Thus any relation between these changes and the course of perturbations to which we may point to, pertains to the domain in which the machine is observed, but not to its organization. Thus, although an autopoietic machine can be treated as an allopoietic machine, this treatment does not reveal its organization as an autopoietic machine.

An organization may remain constant by being static, by maintaining its components constant, or by maintaining constant certain relations between components otherwise in continuous flow or change. Autopoietic machines are organizations of the latter kind: they maintain constant the relations that define them as autopoietic. The actual way in which such an organization may in fact be implemented in the physical space, that is, the physical structure of the machine, varies according to the nature (properties) of the physical materials which embody it. Therefore there may be many different kinds of autopoietic machines in the physical space (physical autopoietic machines); all of them, however, will be organized in such a manner that any physical interference with their operation outside their domain of compensations will result in their disintegration: that is, in the loss of autopoiesis. It also follows that the actual way in which the autopoietic organization is realized in one of these machines (its structure) determines the particular perturbations it can suffer without disintegration, and hence, the domain of interactions in which it can be observed. These features of the actual concreteness of autopoietic machines embodied in physical systems allow us to talk about particular cases, to put them in our domain of manipulation and description, and hence, to observe them in the context of a domain of interactions which is external to their organization. This has two kinds of fundamental consequence:

(i) We can describe physical autopoietic machines, and also manipulate them, as parts of a larger system that defines the independent events which perturb them. Thus, as noted above, we can view these perturbing independent events as inputs, and the changes of the machine that compensate these perturbations as outputs. To do this, however, amounts to treating an autopoietic machine as an allopoietic one, and to recognize that if the independent perturbing events are regular in their nature and occurrence, an autopoietic machine can in fact, be integrated into a larger system as a component allopoietic machine, without any alteration in its autopoietic organization.

(ii) We can analyze a physical autopoietic machine in its physical parts, and treat all its partial homeostatic and regulatory mechanisms as allopoietic machines (sub-machines) by defining their input and output surfaces. Accordingly, these sub-machines are not necessarily components of an autopoietic machine because the relations that define such a machine need not be those that they generate through the input-output relations that define them.

The fact that we can divide physical autopoietic machines into parts does not reveal the nature of the domain of interactions that they define as concrete entities operating in the physical universe.

b. *Living systems*

If living systems are machines, that they are physical autopoietic machines is trivially obvious: they transform matter into themselves in a manner such that the product of their operation is their own organization. However we deem the converse is also true: a physical system if autopoietic, is living. In other words, we claim that the notion of *autopoiesis is necessary and sufficient to characterize the organization of living systems*. This equivalence may not be apparent for some observers due to several reasons which do not pertain to the domain of the organization of autopoietic machines, but which are proper within the domain of description and evaluation of the observers who adopt such reasons, and lead them to its *a priori* negation. The following are some of these reasons:

(i) Machines are generally viewed as human made artifacts with completely known deterministic properties which make them, at least conceptually, perfectly predictable. Contrariwise, living systems are *a priori* frequently viewed as autonomous, ultimately unpredictable systems, with purposeful behavior similar to ours. If living systems were machines, they could be made by man and, according to the view mentioned above, it seems unbelievable that man could manufacture a living system. This view can be easily disqualified, because it either implies the belief that living systems cannot be understood because they are too complex for our meager intellect and will remain so, or that the principles which generate them are intrinsically unknowable; either implication would have to be accepted *a priori* without proper demonstration. There seems to be an intimate fear that the awe with respect to life and the living would disappear if a living system could be not only reproduced, but designed by man. This is nonsense. The beauty of life is not a gift of its inaccessibility to our understanding.

(ii) To the extent that the nature of the living organization is unknown, it is not possible to recognize when one has at hand, either as a concrete synthetic system or as a description, a system that exhibits it. Unless one knows which is the living organization, one cannot know which organization is living. In practice, it is accepted that plants and animals are living but their characterization as living is done through the enumeration of their properties. Among these, reproduction and evolution appear as determinant, and for many observers the condition of living appears subordinated to the possession of these properties. However, when these properties are incorporated in a concrete or conceptual man-made system, those who do not accept emotionally that the nature of life can be understood, immediately conceive of other properties as relevant, and do not accept any synthetic system as living by continuously specifying new requirements.

(iii) It is very often assumed that observation and experimentation are alone sufficient to reveal the nature of living systems and no theoretical analysis is expected to be necessary and least of all sufficient for a characterization of the living organization. It would be long to state why we depart from this radical empiricism. Let us simply say that we believe that epistemological and historical arguments more than justify the contrary view: every experimentation and observation implies a theoretical perspective, and no experimentation or observation has significance or can be interpreted outside the theoretical framework in which it took place.

Our aim was to propose the characterization of living systems that explains the generation of all the phenomena proper to them. We have done this by pointing at autopoiesis in the physical space as a necessary and sufficient condition for a system to be a living one.

To know that a given aim has been attained, is not always easy. In the case at hand, the only possible indication that we have attained our aim is the reader's agreement that all the phenomenology of living systems, including reproduction and evolution, indeed requires and depends on autopoiesis. The following chapters are devoted to show this.

CHAPTER II

DISPENSABILITY OF TELEONOMY

Teleology and teleonomy are notions employed in discourse, descriptive and explanatory, about living systems, and although it is claimed that they do not necessarily enter as causal elements in their functioning, it is asserted that they are essential definitory features of their organization. Our present aim is to show that in the light of the preceding discussion, these notions are unnecessary for the understanding of the living organization.

1. PURPOSELESSNESS

It is usually maintained that the most remarkable feature of living systems is a purposeful organization, or what is the same, the possession of an internal project or program represented and realized in and through their structural organization. Thus, ontogeny is generally considered as an integrated process of development towards an adult state, through which certain structures are attained that allow the organism to perform certain functions according to the innate project which defines it in relation to the environment. Also, phylogeny is viewed as the history of adaptive transformations through reproductive processes aimed at satisfying the project of the species, with complete subordination of the individual to this end. Furthermore, it is apparent that there are organisms that may even appear capable of specifying some purpose in advance (as the authors of this book) and conduct all their activities towards this attainment (heteropoiesis). This element of apparent purpose or the possession of a project or program in the organization of living systems, which has been called teleonomy without implying any vitalistic connotations, is frequently considered as a necessary, if not as a sufficient, definitory feature for their characterization. Purpose or aims, however, as we saw in the first chapter, are not features of the organization of any machine (allo- or autopoietic); these notions belong to the domain of our discourse about our actions, that is, they belong to the domain of descriptions, and when applied to a machine, or any system independent from us, they reflect our considering the machine or system in some encompasssing context. In general, the observer puts the machines either conceptually or concretely to some use, and thus defines a set of circumstances that lead the machine to change, following a certain path of variations in its output. The connection between these outputs, the corresponding inputs, and their relation with the context in which the observer includes them, determine what we call the aim or purpose of the machine; this aim necessarily lies in the domain of the observer that defines the context and establishes the nexuses. Similarly the notion of function arises in the description made by the observer of the components of a machine or system in reference to an encompassing entity, which may be the whole machine or part of it, and whose states constitute the goal that the changes in the components are to bring about. Here again, no matter how direct the causal connections may be between the changes of state of the components and the state which they originate in the total system, the implications in terms of design alluded to by the notion of function are established by the observer and belong exclusively to his domain of description. Accordingly, since the relations implied in the notion of function are not constitutive of the organization of an autopoietic system, they cannot be used to explain its operation.

The organization of a machine, be it autopoietic or allopoietic, only states relations between components and rules for their interactions and transformations, in a manner that specifies the conditions of emergence of the different states of the machine which, then, arise as a necessary outcome whenever such conditions occur. Thus, the notions of purpose and function *have no explanatory value* in the phenomenological domain which they pretend to illuminate, because they do not refer to processes indeed operating in the generation of any of its phenomena. This does not preclude their being adequate for the orientation of the listener towards a given domain of thought. Accordingly, a prediction of a future state of a machine consists only in the accelerated realization in the mind of an observer of its succeeding states, and any reference to an early state to explain a later one in functional or purposeful terms, is an artifice of his description, made in the perspective of his simultaneous mental observation of the two states, that induces in the mind of the listener an abbreviated realization of the machine. Therefore any machine, a part of one or a process that follows a predictable course, can be described by an observer as endowed with a project, a purpose or a function, if properly handled by him with respect to an encompassing context.

Accordingly, if living systems are physical autopoietic machines, teleonomy becomes only an artifice of their description which does not reveal any feature of their organization, but which reveals the consistency in their operation within the domain of observation. Living systems, as physical autopoietic machines, are purposeless systems.

2. INDIVIDUALITY

The elimination of the notion of teleonomy as a defining feature of living systems changes the outlook of the problem completely, and forces us to consider the organization of the individual as the central question for the understanding of the organization of living systems.

In fact, a living system is specified as an individual, as a unitary element of interactions, by its autopoietic organization which determines that any change in it should take place subordinated to its maintenance, and thus sets the boundary conditions that specify what pertains to it and what does not pertain to it in the concreteness of its realization. If the subordination of all changes in a living system to the maintenance of its autopoietic organization did not take place (directly or indirectly), it would lose that aspect of its organization which defines it as a unity, and hence it would disintegrate. Of course it is true for every unity, whichever way it is defined, that the loss of its defining organization results in its disintegration; the peculiarity of living systems, however, is that they disintegrate whenever their autopoietic organization is lost, not that they can disintegrate. As a consequence, all change must occur in each living system without interference with its functioning as a unity in a history of structural change in which the autopoietic organization remains invariant. Thus ontogeny is both an expression of the individuality of living systems and the way through which this individuality is realized. As a process, ontogeny, then, is the expression of the becoming of a system that at each moment is the unity in its fullness, and does not constitute a transit from an incomplete (embryonic) state to a more complete or final one (adult).

The notion of development arises, like the notion of purpose, in the context of observation, and thus belongs to a different domain other than the domain of the autopoietic organization of the living system. Similarly, the conduct of an autopoietic machine that an observer can witness, is the reflection of the paths of changes that it undergoes in the process of maintaining constant its organization through the control of the variables that can be displaced by perturbations, and through the specification in this same process of the values around which these variables are maintained at any moment. Since the autopoietic machine has no inputs or outputs, any correlation between regularly occurring independent events that perturb it, and the state to state transitions that arise from these perturbations, which the observer may pretend to reveal, pertain to the history of the machine in the context of the observation, and not to the operation of its autopoietic organization.

CHAPTER III

EMBODIMENTS OF AUTOPOIESIS

The assertion that physical autopoietic systems are living systems requires the proof that all the phenomenology of a living system can be either reduced or subordinated to its autopoiesis. This proof, obviously, cannot consist in enumerating all biological phenomena and presenting cases of autopoietic systems that exhibit them; rather it must consist in showing that autopoiesis either constitutes or is necessary and sufficient for the occurrence of all biological phenomena, if the proper non-determinant contingencies are given.

1. DESCRIPTIVE AND CAUSAL NOTIONS

An autopoietic system is defined as a unity by its autopoietic organization. The realization of this organization in a physical system requires components which are defined by their role in the autopoiesis and which can only be described in relation to this. Furthermore these components can only be realized by material elements which can exhibit the necessary properties under the conditions specified by the autopoietic organization, and must be produced in the proper topological relation within this organization, by the particular instance (structural realization) of the autopoietic system that they constitute. Accordingly, an autopoietic organization constitutes a closed domain of relations specified only with respect to the autopoietic organization that these relations constitute, and, thus, it defines a 'space' in which it can be realized as a concrete system; a space whose dimensions are the relations of production of the components that realize it:

(i) Relations of constitution that determine that the components produced constitute the topology in which the autopoiesis is realized.

(ii) Relations of specificity that determine that the components produced be the specific ones defined by their participation in the autopoiesis.

(iii) Relations of order that determine that the concatenation of the components in the relations of specification, constitution and order be the ones specified by the autopoiesis.

How these relations of production are embodied in a physical system of course depends on the particular way in which the autopoiesis is realized, that is, on the actual structure of their realization. There are, however, certain general notions which apply to any particular concrete autopoietic system that we must mention at the outset:

(i) Although indeed energetic and thermodynamic considerations would necessarily enter in the analysis of how the components are physically constituted, and in the description of their proper ties in a specific domain of interactions, such that they may satisfy the requirements of their participation in an autopoietic system, these considerations do not enter in the characterization of the autopoietic organization. If the components can be materialized, the organization can be realized; the satisfaction of all thermodynamic and energetic relations is implicit. Thus, for example, in the concrete case of the cell, that we shall consider in the next section, energetic relations that make possible certain reactions with the participation of ATP are not constitutive of the autopoietic organization. However, it is constitutive of the structure through which the autopoietic organization is realized, that the molecules which participate in it should have among their properties the property of entering into the interactions which generate the autopoietic processes and, hence, of holding the required energy relations.

(ii) Notions such as specification and order are referential notions; that is, they do not have meaning outside the context in which they are defined. Thus, when we speak about relations of specification we refer to the specification of components in the context of that which defines the system as autopoietic. Any other element of specificity that may enter, however necessary it may be for the factibility [factual characterization] of the components, but which is not defined through the autopoietic organization, we take for granted. Similarly with the notion of order. Relations of order refer to the establishment of processes that secure the presence of the components in the concatenation that results in autopoiesis. No other reference is meant, however conceivable it may be within other perspectives of description.

(iii) An autopoietic organization acquires topological unity by its embodiment in a concrete autopoietic system which retains its identity as long as it remains autopoietic. Furthermore, the space defined by an autopoietic system is self-contained and cannot be described by using dimensions that define another space. When we refer to our interactions with a concrete autopoietic system, however, we project this system upon the space of our manipulations and make a description of this projection. This we can do because we interact with the components of the autopoietic system through the properties of their constituting elements that do not lie in the autopoietic space, and thus, we modify the structure of the autopoietic system by modifying its components. Our description, however, follows the ensuing change of the projection of the autopoietic system in the space of our description, not in the autopoietic space.

(iv) Notions such as coding and transmission of information do not enter in the realization of a concrete autopoietic system because they do not refer to actual processes in it. Thus, the notion of specificity does not imply coding, information or instructions; it only describes certain relations, determined by and dependent on the autopoietic organization, which result in the production of the specific components. The proper dimension is that of relations of specificity. To say that the system or part of it, codes for specificity, is not only a misnomer but also misleading; this is so, because such an expression represents a mapping of a process that occurs in the space of autopoiesis onto a process that occurs in the space of human design (heteropoiesis), and it is not a reformulation of the phenomenon. The notion of coding is a cognitive notion which represents the interactions of the observer, not a phenomenon operative in the observed domain. The same applies to the notion of regulation. This notion is valid in the domain of description of heteropoiesis, and it reflects the simultaneous observation and description made by the designer (or his equivalent) of interdependent transitions of the system that occur in a specified order and at specified speeds. The corresponding dimension in an autopoietic system is that of relations of production of order, but here again only in the context of the autopoiesis and not of any particular state of the system as it would appear projected on our domain of descriptions. The notion of regulation, then, can enter in the description, but does not refer to an actual process in the autopoietic organization.

2. MOLECULAR EMBODIMENTS

That a cell is an autopoietic system is trivially apparent in its life cycle. What is not trivial is how the cell is a molecular embodiment of autopoiesis, as it should be apparent in its analysis in terms of the dimensions of its autopoietic space:

(i) *Production of Constitutive Relations*

Constitutive relations are relations that determine the topology of the autopoietic organization, and hence its physical boundaries. The production of constitutive relations through the production of the components that hold these relations is one of the defining dimensions of an autopoietic system. In the cell such constitutive relations are established through the production of molecules (proteins, lipids, carbohydrates and nucleic acids) which determine the topology of the relations of production in general; that is, molecules which determine the relations of physical neighborhood necessary for the components to hold the relations that define them. The cell defines its physical boundaries through its dimension of production of constitutive relations that specify its topology. There is no specification in the cell of what it is not.

(ii) *Production of Relations of Specifications*

Relations of specifications are relations that determine the identity (properties) of the components of the autopoietic organization, and hence, in the case of the cells, its physical factibility. The establishment of relations of specification through the production of components that can hold these relations is another of the defining dimensions of an autopoietic system. In the cell such relations of specification are produced mainly through the production of nucleic acids and proteins that determine the identity of the relations of production in general. In the cell this is obviously obtained, on the one hand, by relations of specificity between DNA, RNA and proteins, and on the other hand, by relations of specificity between enzymes and substrates. Such production of relations of specification holds only within the topological substrate defined by the production of relations of constitution. There is no production in the cell as an autopoietic system of relations of specification that do not pertain to it.

(iii) *Production of Relations of Order*

Relations of order are those that determine the dynamics of the autopoietic organization by determining the concatenation of the production of relations of constitution, specification and order, and hence its actual realization. The establishment of relations of order through the production of components that realize the production of relations of constitution, specification and order, constitute the third dimension of the autopoietic space. In the cell, relations of order are established mainly by the production of components (metabolites, nucleic acids and proteins) that control the speed of production of relations of constitution, specification and order. Relations of order, thus, conform a network of parallel and sequential relations of constitution, specification and order that constitute the cell as a system in which the relations of production that specify this network as a dynamic physical topological unity, are maintained constant. There is no ordering through the autopoietic organization of the cell of processes that do not belong to it.

If one examines a cell it is apparent that:
DNA participates in the specification of polypeptides, and hence, or proteins, enzymatic and structural, which specifically participate in the production of proteins, nucleic acids, lipids, glucides and metabolites. Metabolites (which include all small molecules, monomers or not, produced in the cell) participate in the determination of the speed of the various processes and reactions that constitute the cell, establishing a network of interrelated speeds in parallel and sequentially interconnected processes, both by gating and by constitutive participation, in a way such that every reaction is a function of the state of the transforming network that they integrate. All processes occur bound to a topology determined by their participation in the processes of production of relations of constitution.

We as observers can project all cellular processes upon a system of three orthogonal coordinates, and legitimately say, as valid in the projection, that specification is mostly produced by nucleic acids, constitution by proteins, and order (regulation) by metabolites. The autopoietic space, however, is curved and closed in the sense that it is entirely specified by itself, and such a projection represents our cognitive relation with it, but does not reproduce it. In it, specification takes place at all points where its organization determines a specific process (protein synthesis, enzymatic action, selective permeability); ordering takes place at all points where two or more processes meet (changes of speed or sequence, allosteric effects, competitive and non-competitive inhibition, facilitation, inactivation, etc.) determined by the structure of the participating components; constitution occurs at all places where the structure of the components determines physical neighborhood relations (membranes, particles, active site in enzymes). What makes this system a unity with identity and individuality is that all the relations of production are coordinated in a system describable as an homeostatic system that has its own unitary character as the variable that it maintains constant through the production of its components. In such a system any deformation at any place is not compensated by bringing the system back to an identical state of its components as it would be described by projecting it upon a three-dimensional Cartesian space; rather it is compensated by keeping its organization constant as defined by the relation of the relations of production of relations of constitution, specification and order which constitutes autopoiesis. In other words, compensation of deformation keeps the autopoietic system in the autopoietic space.

That all the biological features of the cell as a unity are determined by its autopoiesis, is henceforth obvious. In fact, the only thing that defines the cell as a unity (as an individual) is its autopoiesis, and thus the only restriction placed on the existence of the cell is the maintenance of autopoiesis. All the rest – that is, its structure – can vary: relations of topology, specificity and order can vary as long as they constitute a network in an autopoietic space.

3. ORIGIN

The production of relations of constitution, specification and order, are not exclusive to autopoietic systems. They are inherent to unitary interactions in general, and to molecular interactions in particular; they depend on the properties of the units or molecules as expressed in the geometric and energetic relationships which they may adopt. Thus, the geometric properties of the molecules determine the relations of constitution, that is, the topology, the physical neighborhoods or spatial relations in which they may enter. The chemical properties of the molecules determine their possible interactions, and, hence, the relations of specificity which are a dimension orthogonal to relations of constitution. Both together, they determine sequence and concatenation of molecular interactions, that is, relations of order. Accordingly, autopoiesis may arise in a molecular system if the relations of production are concatenated in such a way that they produce components that specify the system as a unity which exists only while it is actively produced by such concatenation of processes. This is to say that autopoiesis arises in a molecular system only when the relation that concatenates these relations is produced and maintained constant through the production of the molecular components that constitute the system through this concatenation. Thus, in general, the question of the origin of an autopoietic system is a question about the conditions that must be satisfied for the establishment of an autopoietic space. This problem, then, is not a chemical one, in terms of what molecules took or can take part in the process, but a general one of what relations the molecules or any constitutive units should satisfy. This deserves the following considerations:

(i) An autopoietic system is defined as a unity by and through its autopoietic organization. This unity is, thus, a topological unity in the space in which the components have existence as entities that may interact and have relations. For living systems such a space is the physical space. Without unity in some space an autopoietic system is not different from the background in which it is supposed to lie, and, hence, can only be a system in the space of our description where its unity is conceptually stipulated. Without unity in the physical space a living system would lack the dynamics of production relations which constitute it as a concrete entity in that space.

(ii) The establishment of an autopoietic system cannot be a gradual process; either a system is an autopoietic system or it is not. In fact, its establishment cannot be a gradual process because an autopoietic system is defined as a system; that is, it is defined as a topological unity by its organization. Thus, either a topological unity is formed through its autopoietic organization, and the autopoietic system is there and remains, or there is no topological unity, or a topological unity is formed in a different manner and there is no autopoietic system but there is something else. Accordingly, there are not and there cannot be intermediate systems. We can describe a system and talk about it as if it were a system which, with a little transformation, would become an autopoietic system because we can imagine different systems with which we compare it, but such a system would be intermediate only in our description, and in no organizational sense would it be a transition system.

(iii) Autocatalytic processes do not constitute autopoietic systems because among other things, they do not determine their topology. Their topology is determined by a container that is part of the specification of the system, but which is independent of the operation of the autocatalysis. Processes of this or similar kind are abundant in the physical space. Coupling of independent processes into larger systems is also the rule; these may or may not constitute unities defined by the circumstances of their constitution in a given space, be this space physical or otherwise. They, however, will not constitute or participate in the constitution of an autopoietic system unless the system they conform becomes defined as a topological unity through its embodiment of an autopoietic organization. A unity is defined by an operation of distinction; in an autopoietic system its autopoiesis constitutes the operation of distinction that defines it, and its origin is cocircumstantial with the establishment of this operation.

(iv) The problem of the origin of autopoietic systems has two aspects; one refers to their factibility, and the other to the possibility of their spontaneous occurrence. The first aspect can be stated in the following manner:

the establishment of any system depends on the presence of the components that constitute it, and on the kinds of interactions in which they may enter; thus, given the proper components and the proper concatenation of their interactions, the system is realized. The concrete question about the factibility of a molecular autopoietic system is, then, the question of the conditions in which different chemical processes can be concatenated to form topological unities that constitute relational networks in the autopoietic space. The second aspect can be stated in the following manner: given the factibility of autopoietic systems, and given the existence of terrestrial autopoietic systems, there are natural conditions under which these may be spontaneously generated. Concretely the question would be, 'What were or are the natural conditions under which the components of the autopoietic systems arose or arise spontaneously on the earth, and concatenate to form them?' This question cannot be answered independently of the manner in which the factibility question is answered, particularly in what refers to the factibility of one or several different kinds of molecular autopoietic systems. The presence today of one mode of autopoietic organization on the earth (the nucleic acid protein system), cannot be taken to imply that the factibility question has only one answer.

The notions that we have discussed are valid for the origin (constitution) of autopoietic systems at any level of physical embodiment, molecular or supramolecular. We shall not dwell on the particular circumstances of the establishment of any of these embodiments. We shall leave this matter for another inquiry, accepting the existence of living systems as an existential proof of the factibility of the spontaneous generation of autopoietic systems. We shall consider next the significance of the conditions of topological unity for the diversity of autopoietic systems.

CHAPTER IV

DIVERSITY OF AUTOPOIESIS

Living systems embody the living organization. Living systems are autopoietic systems in the physical space. The diversity of living systems is apparent; it is also apparent that this diversity depends on reproduction and evolution. Yet, reproduction and evolution do not enter into the characterization of the living organization, and living systems are defined as unities by their autopoiesis. This is significant because it makes the phenomenology of living systems dependent on their being autopoietic unities. In fact, reproduction requires the existence of a unity to be reproduced, and it is necessarily secondary to the establishment of such a unity; evolution requires reproduction and the possibility of change, through reproduction of that which evolves, and it is necessarily secondary to the establishment of reproduction. It follows that the proper evaluation of the phenomenology of living systems, including reproduction and evolution, requires their proper evaluation as autopoietic unities.

1. SUBORDINATION TO THE CONDITION OF UNITY

Unity (distinguishability from a background, and, hence, from other unities), is the sole necessary condition for existence in any given domain. In fact, the nature of a unity and the domain in which it exists are specified by the process of its distinction and determination; this is so regardless of whether this process is conceptual (as when a unity is defined by an observer through an operation of distinction in his domain of discourse and description), or whether this process is physical (as when a unity becomes established through the actual working of its defining properties that assert its distinction from a background through their actual operation in the physical space). Accordingly, different kinds of unities necessarily differ in the domain in which they are established, and having different domains of existence they may or may not interact according to whether these domains do or do not intersect. Unity distinction [the distinctiveness and distinguishing of unity], then, is not an abstract notion of purely conceptual validity for descriptive or analytical purposes, but it is an operative notion referring to the process through which a unity becomes asserted or defined: the conditions which specify a unity determine its phenomenology. In living systems, these conditions are determined by their autopoietic organization. In fact, autopoiesis implies the subordination of all change in the autopoietic system to the maintenance of its autopoietic organization, and since this organization defines it as a unity, it implies total subordination of the phenomenology of the system to the maintenance of its unity. This subordination has the following consequences:

(i) The establishment of a unity defines the domain of its phenomenology, but given the way the unity is constituted by its structure defines the kind of phenomenology that it generates in that domain. It follows that the particular form adopted by the phenomenology of each autopoietic (biological) unity depends on the particular way in which its individual autopoiesis is realized. It also follows that the domain of ontogenic transformations (including conduct) of each individual is the domain of the homeostatic trajectories through which it can maintain its autopoiesis.

(ii) All the biological phenomenology is necessarily determined and realized through individual autopoietic unities in the physical space, and consists of all the paths of transformations that they undergo as homeostatic systems, singly or in groups, in the process of maintaining constant their defining individual relations. Whether in the process of their interactions the autopoietic unities do or do not unite to constitute additional unities, is irrelevant for the subordination of the biological phenomenology to the maintenance of the identity of the individual unities. If united they produce a new unity that is not autopoietic, its phenomenology, that will necessarily depend on its organization, will be biological or not according to its dependence on the autopoiesis of its components, and will accordingly depend or not on the maintenance of these as autopoietic units. If the new unity is autopoietic, its phenomenology is directly biological and obviously depends on the maintenance of its autopoiesis, which in turn may or may not depend on the autopoiesis of its components.

(iii) The identity of an autopoietic unity is maintained as long as it remains autopoietic; that is, as long as it, as a unity in the physical space, remains a unity in the autopoietic space, regardless of how much it may otherwise be transformed in the process of maintaining its autopoiesis.

(iv) Only after a unity has been constituted as an autopoietic unity (individual) can reproduction take place as a biological phenomenon.

2. PLASTICITY OF ONTOGENY

Ontogeny is the history of the structural transformation of a unity. Accordingly, the ontogeny of a living system is the history of maintenance of its identity through continuous autopoiesis in the physical space. From the mere fact that a physical autopoietic system is a dynamic system, realized through relations of productions of components that imply concrete physical interactions and transformations, it is a necessary consequence of the autopoietic organization of a living system that its ontogeny should take place in the physical space. There are several comments to this notion of ontogeny:

(i) Since the way an autopoietic system maintains its identity depends on its particular way of being autopoietic, that is, on its particular structure, different classes of autopoietic systems have different classes of ontogenies.

(ii) Since an autopoietic system does not have inputs or outputs, all the changes that it may undergo without loss of identity, and, hence, with maintenance of its defining relations, are necessarily determined by its homeostatic organization. Consequently, the phenomenology of an autopoietic system is necessarily always commensurate with the deformations that it suffers without loss of identity, and with the deforming ambience in which it lies. Otherwise it would disintegrate.

(iii) As a consequence of the homeostatic nature of the autopoietic organization, the way the autopoiesis is realized in any given unity may change during its ontogeny, with the sole restriction that this should take place without loss of identity, that is, through uninterrupted autopoiesis.

(iv) Although the changes that an autopoietic system may undergo without loss of identity while compensating its deformations under interactions are determined by its organization, the sequence of such changes is determined by the sequence of these deformations. There are two sources of deformations for an autopoietic system as they appear to be to an observer: one is constituted by the external environment as a source of independent events in the sense that these are not determined by the organization of the system; the other is constituted by the system itself as a source of states which arise from compensations of deformations, but which themselves can constitute deformations that generate further compensatory changes. In the phenomenology of the autopoietic organization these two sources of perturbations are indistinguishable, and in each autopoietic system they braid together to form a single ontogeny. Thus, although in an autopoietic system all changes are internally determined, for an observer its ontogeny partly reflects its history of interactions with an independent ambience. Accordingly, two otherwise equivalent autopoietic systems may have different ontogenies.

(v) An observer beholding an autopoietic system as a unity in a context that he also observes, and which he describes as its environment, may distinguish in it internally and externally generated perturbations, even though these are intrinsically indistinguishable for the autopoietic system itself. The observer can use these distinctions to make statements about the history of the autopoietic system which he observes, and he can use this history to describe an ambience (which he infers) as the domain in which the system exists. He cannot, however, infer from the observed correspondence between the ontogeny of the system and the ambience which this ontogeny describes, or from the environment in which he sees it, a constitutive representation of these in the organization of the autopoietic systems. The continuous correspondence between conduct and ambience revealed during ontogeny is the result of the homeostatic nature of the autopoietic organization, and not of the existence of any representation of the ambience in it; nor is it at all necessary that the autopoietic system should obtain or develop such a representation to persist in a changing ambience. To talk about a representation of the ambience, or the environment, in the organization of a living system may be metaphorically useful, but it is inadequate and misleading to reveal the organization of an autopoietic system.

(vi) The compensatory changes that an autopoietic system may undergo while retaining its identity, may be of two possible kinds according to how its structure is affected by the perturbations: they may be (a) conservative changes in which only the relations between the components change; or they may be (b) innovative changes in which the components themselves change. In the first case, the internal or external interactions causing the deformations do not lead to any change in the way the autopoiesis is realized, and the system remains in the same point in the autopoietic space because its components are invariant; in the second case, on the contrary, the interactions lead to a change in the way the autopoiesis is realized and, hence, to a displacement of the system in the autopoietic space because its components changed. Accordingly, while the first case implies a conservative ontogeny, the second case implies an ontogeny which is also a process of specification of a particular autopoiesis that in its determination is, necessarily, a function of both the plasticity of the components of the system and the history of its interactions.

3. REPRODUCTION, A COMPLICATION OF THE UNITY

Reproduction requires a unity to be reproduced; this is why reproduction is operationally secondary to the establishment of the unity, and it cannot enter as a defining feature of the organization of living systems. Furthermore, since living systems are characterized by their autopoietic organization, reproduction must necessarily have arisen as a complication of autopoiesis during autopoiesis, and its origin must be viewed and understood as secondary to, and independent from the origin of the living organization. The dependence of reproduction upon the existence of the unity to be reproduced is not a trivial problem of precedence, but it is an operational problem in the origin of the reproduced system and its relations with the reproducing mechanism. Accordingly, in order to understand reproduction and its consequences in autopoietic systems we must analyze the operational nature of this process in relation to autopoiesis.

(i) There are three phenomena that must be distinguished in relation to the notion of reproduction; these are replication, copy and self-reproduction. *Replication*. A system which successively generates unities different from itself, but in principle identical to each other, and with an organization which the system determines in the process of their production, is a replicating system. Replication, then, is not different from repetitive production. Any distinction between these processes arises as a matter of description in the emphasis that the observer puts on the origin of the equivalent organization of the successively produced unities, and on the relevance that this equivalence has in a domain different from that in which the repetitive production takes place. Thus, although all molecules are produced by specific molecular and atomic processes that can at least in principle be repeated, only when certain specific kinds of molecules are produced in relation to the cellular activities (proteins and nucleic acids) by certain repeatable molecular concatenations is their production called replication. Such a denomination then, strictly, makes reference only to the context in which the identity of the successively produced molecules is deemed necessary, not to a unique feature of that particular molecular synthesis.

Copy. Copy takes place whenever a given object or phenomenon is mapped by means of some procedure upon a different system, so that an isomorphic object or phenomenon is realized in it. In the notion of copy the emphasis is put on the mapping process, regardless of how this is realized, even if the mapping operation is performed by the model unit itself.

Self-reproduction. Self-reproduction takes place when a unity produces another one with a similar organization to its own, through a process that is coupled to the process of its own production. It is apparent that only autopoietic systems can self-reproduce because only they are realized through a process of self-production (autopoiesis).

(ii) For an observer there is reproduction in all these three processes because he can recognize in each of them a unitary pattern of organization which is embodied in successively generated systems through the three well defined mechanisms. The three processes, however, are intrinsically different because their dynamics give rise to different phenomenologies which appear particularly distinct if one considers the network of systems generated under conditions in which change is allowed in the process of reproduction of the successively embodied pattern of organization. Thus, in replication and copy the mechanism of reproduction is necessarily external to the pattern reproduced, while in self-reproduction it is necessarily identical to it. Furthermore, only in self-copy and self-reproduction can changes in the unities produced which embody the pattern reproduced affect the reproducing mechanism. The consequences of this will be dealt with in the next section, but now it should be clear that the historical interconnections established between independent unities through reproduction varies with the mechanism through which reproduction is achieved.

(iii) In living systems presently known on earth autopoiesis and reproduction are directly coupled and, hence, these systems are truly self-reproducing systems. In fact, in them reproduction is a moment in autopoiesis, and the same mechanism that constitutes one constitutes the other. The consequences of such a coupling are paramount: (a) Self-reproduction must take place during autopoiesis. Accordingly the network of individuals thus produced is necessarily self-contained in the sense that it does not require for its establishment a mechanism independent of the autopoietic determination of the self-reproducing unities. Such would not be the case if reproduction were attained through external copy or replication. (b) Self-reproduction is a form of autopoiesis; therefore, variation and constancy in each reproductive step are not independent and both must occur as expressions of autopoiesis. (c) Variation through self-reproduction of the way the autopoiesis is realized can only arise as a modification during autopoiesis of a pre-existing functioning autopoietic structure; consequently, variation through self-reproduction can only arise from perturbations that require further homeostatic complications to maintain autopoiesis constant. The history of self-reproductively connected autopoietic systems can only be one of continuous complication of autopoiesis.

(iv) The nature of reproduction depends on the nature of the unity. The same goes for its origin. Replication takes place independently of autopoiesis. Copy takes place only in heteropoiesis, and can be deemed to take place in other situations solely as a description. Self-reproduction is exclusively associated to autopoiesis and its origin is bound to it as a historically secondary phenomenon. The reason for this association will be dealt with in the next section.

(v) Notions such as coding, message or information are not applicable to the phenomenon of self-reproduction; their use in the description of this phenomenon constitutes an attempt to represent it in the language of heteropoietic design. In fact, the notions of coding, message and transmission of information apply only to the reduction of uncertainties in the communicative interactions between independent unities under conditions in which the messenger acts as an arbitrary non-participant link. Nucleic acids are constitutive components in the process of autopoiesis, not arbitrary links between independent entities. Thus, in self-reproduction there is no transmission of information between independent entities; the reproducing and the reproduced unities are topologically independent entities produced through a single process of autopoiesis in which all components have a constitutive participation.

4. EVOLUTION, A HISTORICAL NETWORK

A historical phenomenon is a process of change in which each state of the successive states of a changing system arises as a modification of a previous state in a causal transformation, and not *de novo* as an independent occurrence. Accordingly, the notion of history may either be used to refer to the antecedents of a given phenomenon as the succession of events that gave rise to it, or it may be used to characterize the given phenomenon as a process. Therefore, since an explanation is always given in the present as a reformulation of the phenomenon to be explained in the domain of interactions of its components (or of isomorphic elements), the history of a phenomenon as a description of its antecedents cannot contribute to its explanation because the antecedents are not components of the phenomenon which they precede or generate. Conversely, since history as a phenomenon is to be explained in the present as a changing network of sequentially produced events in which each event as a state of the network arises in it as a transformation of the previous state, it follows that although history cannot contribute to explain any phenomenon, it can permit an observer to account for the origin of a phenomenon as a present state in a changing network. This he can do because he has observational (or descriptive) independent access to the different states of the historical process. It is in this context that the phenomenology of autopoietic systems must be considered when viewed in reference to evolution. Biological evolution is a historical phenomenon and as such it must be explained in the present by its reformulation as a historical network constituted through the causal interactions of coupled or independent biological events. Furthermore, biological events depend on the autopoiesis of living systems; accordingly, our aim here is to understand how evolution is defined as a historical process by the autopoiesis of the biological unities.

(i) If by evolution we refer to what has taken place in the history of transformation of terrestrial living systems, evolution is the history of change in the realization of an invariant organization embodied in independent unities sequentially generated through reproductive steps, in which the particular structural realization of each unity arises as a modification of the preceding one (or ones) which, thus, constitutes both its sequential and historical antecedent. Consequently, evolution requires sequential reproduction and change in each reproductive step. Without sequential reproduction as a reproductive process in which the structural realization of each unity in the sequence constitutes the antecedent for the structural realization of the next one, there is no history; without change in each sequential reproductive step, there is no evolution. In fact, sequential transformations in a unity without change of identity constitute its ontogeny, that is, its individual history if it is an autopoietic unity.

(ii) Reproduction by replication or copy of a single unchanging model implies an intrinsic uncoupling between the organization of the unities produced and their producing mechanism. As a consequence, any change in the realization of the organization embodied in the unities successively produced by replication or copy from a single model, can only reflect the ontogenies of the reproducing systems or the independent ontogenies of the units themselves. The result is that under no circumstance in these non-sequential reproductive cases does a change in the structure of a unity affect the structure of the others yet to be produced, and, independently of whether they are autopoietic or not, they do not constitute a historical network, and no evolution takes place. The collection of unities thus produced constitutes a collection of independent ontogenies. In sequential reproduction, as it occurs in self-reproducing systems which attain reproduction through autopoiesis, or as it occurs in those copying systems in which each new unity produced constitutes the model for the next one, the converse is true. In these cases, there are aspects of the structural realization of each unity that determine the structure of the next one by their direct coupling with the reproductive process which is, thus, subordinated to the organization of the reproduced unities. Consequently, changes in these aspects of the structure of the unities sequentially generated, that occur either during their own ontogeny or in the process of their generation, necessarily result in the production of an historical network in which the unities successively produced embody an invariant organization in a changing structure as each unity arises as a modification of the previous one. In general, then, sequential reproduction with the possibility of change in each reproductive step necessarily leads to evolution, and in particular, in autopoietic systems evolution is a consequence of self-reproduction.

(iii) Ontogeny and evolution are completely different phenomena, both in their outlook and in their consequences. In ontogeny, as the history of transformation of a unity, the identity of the unity, in whatever space it may exist, is never interrupted. In evolution, as a process of historical change there is a succession of identities generated through sequential reproduction which constitute a historical network, and that which changes (evolves), the pattern of realization of the successively generated unities exists in a different domain than the unities that embody it. A collection of successive ontogenies in whose structure an observer can see relations of maintained change, but which have not been generated through sequential reproduction, do not constitute an evolving system, not even if they reflect the continuous transformation (ontogeny) of the system that produced them. It is inadequate to talk about evolution in the history of change of a single unity in whatever space it may exist; unities only have ontogenies. Thus, it is inadequate to talk about the evolution of the universe, or the chemical evolution of the earth; one should only talk about the ontogeny of the universe or the chemical history of the earth. Also, there is biological evolution only since there is sequential reproduction of living systems; if there were non self-reproducing autopoietic systems before that, their different patterns of realization did not evolve, and there was only the history of their independent ontogenies.

(iv) Selection, as a process in a population of unities, is a process of differential realization in a context that specifies the unitary structures that can be realized. In a population of autopoietic unities selection is a process of differential realization of autopoiesis, and, hence, if these are self-reproducing autopoietic unities, of differential self-reproduction. Consequently, if there is sequential reproduction, and the possibility of change in each reproductive step, selection can make the transformation of the reproducible structural patterns realized in each successive unity a recursive function of the domain of interactions which that very same autopoietic unity specifies. If any system that is realized is necessarily adapted in the domain in which it is realized, and adaptation is the condition of possible realization for any system, evolution takes place only if adaptation is conserved by the unities that embody the invariant organization of the evolving lineage. Accordingly, different evolving systems would differ only in the domain in which they are realized, and, hence, in which selection takes place, not in whether they are adaptive or not. Thus, evolution in self-reproducing living systems that maintain their identity in the physical space (while the realization of their autopoietic organization is commensurate with the restrictions of the ambience in which they exist), is necessarily a process of continued adaptation because only those of them whose autopoiesis can be realized reproduce, regardless of how much the way they are autopoietic may otherwise change in each reproductive step.

(v) For evolution to take place as a history of change in the realization of an invariant organization embodied in successively generated unities, reproduction must allow for structural change in the sequentially reproduced unities. In present living systems reproduction takes place as a modification of autopoiesis and is bound to it. This was to be expected. Originally many kinds of autopoietic unities were probably formed which would mutually compete for the precursors. If any class of them had any possibility of self-reproduction, it is evident that it would immediately displace through selection the other non-reproducing forms. The onset of the history of self-reproduction need not have been complex; for example, in a system with distributed autopoiesis mechanical fragmentation is a form of self-reproduction. Evolution through selection would appear with the enhancement of those features of the autopoietic unities that facilitated their fragmentation (and hence the regularity and frequency of self-reproduction) to the extent of making it independent of external accidental forces. Once the most simple self-reproducing process takes place in an autopoietic system, evolution is on its course and self-reproduction can enter in a history of change, with the ensuing total displacement of any co-existing non-self-reproducing autopoietic unities. *Hence the linkage between autopoiesis and self-reproduction in terrestrial living systems*. Of course it is not possible to say now what actually took place in the origin of biological evolution, but this does not seem to offer an insurmountable conceptual difficulty. The fact is that in present day living systems self-reproduction is crucially associated to nucleic acids and their role in protein specification. We think that this could not have been so if the nucleic acid-protein association were not a condition virtually constitutive of the original autopoietic process which was secondarily associated to reproduction and variation; and we think that this is so because only uncompensated changes at the level of the autopoietic process itself can be incorporated (through sequential reproduction) as reproducible changes of the autopoietic organizations of the next unity in a manner that allows for evolution to take place. What is not apparent, though, is whether or not there have been other modes of autopoietic realization, and other sources of variation, than those associated with the nucleic acid-protein system, in the history of terrestrial living systems. Whichever the case, once self-reproduction appears in autopoiesis, any perturbation which modifies the way in which the autopoiesis is realized, can, in principle, be reproduced in the next generation, and, thus, be the source of variations if the change affected those processes involved in reproduction. Accordingly the phenomenology of biological evolution and its origin rests on the inception of two processes: self-reproduction and variation. One refers to possible forms of complication of the autopoiesis, the other to the introduction of perturbations which irreversibly modify the way the autopoiesis is realized. Both undergo historical transformations, which, though coupled, are not equivalent.

(vi) Of the two possible mechanisms that can give rise to sequential reproduction, the only one which is accessible to autopoietic systems in the absence of an independent copying mechanism is self-reproduction, because of the coincidence between the reproducing mechanisms and the reproducing unity. Sequential reproduction through copy takes place at present only in relation to the operation of living systems in their domain of interactions, particularly in cultural learning; cultural evolution takes place through sequential copy of a changing model in the process of social indoctrination generation after generation.

(vii) A species is a population or collection of populations of reproductively interconnected individuals which are thus nodes in a historical network. Genetically these individuals share a genetic pool, that is, a fundamentally equivalent pattern of autopoietic realization under historical transformations. Historically, a species arises when a reproductive network of this kind develops an independent reproductive network as a branch which by being an independent historical network (reproductively separated) has an independent history. It is said that what evolves is the species and that the individuals in their historical existence are subordinated to this evolution. In a superficial descriptive sense this is meaningful because a particular species as an existing collection of individuals represents continuously the state of a particular historical network in its process of becoming one, and, if described as a state of a historical network, a species necessarily appears in a process of transformation. Yet, the species exists as a unity only in the historical domain, while the individuals that constitute the nodes of the historical network exist in the physical space. Strictly, a historical network is defined by each and every one of the individuals which constitute its nodes, but it is at any moment represented historically by the species as the collection of all the simultaneously existing nodes of the network; in fact, then, a species does not evolve because as a unity in the historical domain it only has a history of change. What evolves is a pattern of autopoietic realization embodied in many particular variations in a collection of transitory individuals that together define a reproductive historical network. Thus, the individuals, though transitory, are essential, not dispensable, because they constitute a necessary condition for the existence of the historical network which they define. The species is only an abstract entity in the present, and although it represents a historical phenomenon it does not constitute a generative factor in the phenomenology of evolution, it is its result.

5. SECOND AND THIRD ORDER AUTOPOIETIC SYSTEMS

Whenever the conduct of two or more unities is such that there is a domain in which the conduct of each one is a function of the conduct of the others, it is said that they are coupled in that domain. Coupling arises as a result of the mutual modifications that interacting unities undergo in the course of their interactions without loss of identity. If the identity of the interacting unities is lost in the course of their interactions, a new unity may be generated as a result of it, but no coupling takes place. In general, however, coupling leads also to the generation of a new unity that may exist in a different domain from the domain in which the component-coupled unities retain their identity. The way in which this takes place, as well as the domain in which the new unity is realized, depend on the properties of the component unities. Coupling in living systems is a frequent occurrence; the following comments are meant to show that the nature of the coupling of living systems is determined by their autopoietic organization.

(i) Autopoietic systems can interact with each other without loss of identity as long as their respective paths of autopoiesis constitute reciprocal sources of compensable disturbances. Furthermore, due to their homeostatic organization autopoietic systems can couple and constitute a new unity while their individual paths of autopoiesis become reciprocal sources of specification of each other's ambience, if their reciprocal deformations do not overstep their corresponding ranges of tolerance for variation without loss of autopoiesis. As a consequence the coupling remains invariant while the coupled systems undergo structural changes selected through the coupling and, hence, commensurate with it. These considerations also apply to the coupling of autopoietic and non-autopoietic unities with obvious modifications in relation to the retention of identity of the latter. In general, then, the coupling of autopoietic systems with other unities, autopoietic or not, is realized through their autopoiesis. That coupling may facilitate autopoiesis requires no further discussion, and that this facilitation may take place through the particular way in which the autopoiesis of the coupled unities is realized has already been said. It follows that selection for coupling is possible, and that through evolution under a selective pressure for coupling a composite system can be developed (evolved) in which the individual autopoiesis of every one of its autopoietic components is subordinated to an ambience defined through the autopoiesis of all the other autopoietic components of the composite unity. Such a composite system will necessarily be defined as a unity by the coupling relations of its component autopoietic systems in a space that the nature of the coupling specifies, and will remain as a unity as long as the component systems retain their autopoiesis which allows them to enter into those coupling relations.

A system generated through the coupling of autopoietic unities may, on a first approximation, be seen by an observer as autopoietic to the extent that its realization depends on the autopoiesis of the unities which integrate it. Yet, if such a system is not defined by relations of production of components that generate these relations and define it as a unity in a given space, but by other relations, either between components or processes, it is not an autopoietic system and the observer is mistaken. The apparent autopoiesis of such a system is incidental to the autopoiesis of the coupled unities which constitute it, and not intrinsic to its organization; the mistake of the observer, therefore, lies in that he sees the system of coupled autopoietic unities as a unity in his perceptive domain in terms other than those defined by its organization. In contrast, a system realized through the coupling of autopoietic unities and defined by relations of production of components that generate these relations and constitute it as a unity in some space, is an autopoietic system in that space regardless of whether the components produced coincide or not with the unities which generate it through their coupled autopoiesis. If the autopoietic system thus generated is a unity in the physical space it is a living system. An autopoietic system whose autopoiesis entails the autopoiesis of the coupled autopoietic unities which realize it, is an autopoietic system of higher order.

In general, the actual recognition of an autopoietic system poses a cognitive problem that has both to do with the capacity of the observer to recognize the relations that define the system as a unity, and with his capacity to distinguish the boundaries which delimit this unity in the space in which it is realized. Since it is a defining feature of an autopoietic system that it should specify its own boundaries, a proper recognition of an autopoietic system as a unity requires that the observer performs an operation of distinction that defines the limits of the system in the same domain in which it specifies them through its autopoiesis. If this is not the case he does not observe the autopoietic system as a unity, even though he may conceive it. Thus, presently, the recognition of a cell as a molecular autopoietic unity offers no serious difficulty because we can identify the autopoietic nature of its organization, and interact visually, mechanically and chemically, with one of the boundaries (membrane) which its autopoiesis generates as an interface which delimits it as a three dimensional physical unity. In addition the observer may have two kinds of difficulties in the identification of an autopoietic unity as an actually distinguishable system: on the one hand, he may treat the system as a unity by making an operation of distinction in a space different from the space in which it is realized because he has not yet properly recognized the relations of production of components that constitute it, and, hence, cannot recognize the topological relations which specify its unity in that space; on the other hand, due to his own mode of autopoietic organization (and, hence, cognitive structure) he may be unable to interact in the space in which the system is realized as a unity, and, hence, he may be unable to observe it as a unity because he cannot specify the proper perceptual dimensions. In the first case, the observer makes a unity distinction which is not commensurate with the autopoietic system, and he thus defines and operates with a different unity; in the second case he makes no distinction at all, and he has no unity with which to operate. In either case the phenomenology of the autopoietic unity remains unobservable. However, if there is no misidentification of the system, even if its unity is not yet operationally observable, its phenomenology can be asserted by the recognition of the organization that constitutes it.

(ii) An autopoietic system can become a component of another system if some aspects of its path of autopoietic change can participate in the realization of this other system. As has been said, this can take place in the present through a coupling that makes use of the homeostatic resorts of the interacting systems, or through evolution by the recursive effect of a maintained selective pressure on the course of transformation of a reproductive historical network, which results in a subordination of the individual component autopoiesis (through historical change in the way these are realized) to the ambience of reciprocal perturbations which they specify. Whichever the case, an observer can describe an autopoietic component of a composite system as playing an allopoietic role in the realization of the larger system which it contributes to realize through its autopoiesis. In other words, the autopoietic unity functions in the context of the composite system in a manner that the observer would describe as allopoietic. Yet, the allopoietic function is exclusively a feature of the description and pertains to a frame of reference defined by the observer. As we described in Chapter I, there are allopoietic machines whose organization is intrinsically different from autopoietic machines, and can be described (with no reference to function) by pointing out that the product of their operation is different from themselves. Accordingly, when an autopoietic system is described as having an allopoietic role as a component in a larger system, the description makes reference only to its participation in the production of relations that adopt the form proper to an allopoietic system, but nothing is implied about function which is proper only in the domain of heteropoietic human design.

(iii) If the autopoiesis of the component unities of a composite autopoietic system conforms to allopoietic roles that through the production of relations of constitution, specification and order define an autopoietic space, the new system becomes in its own right an autopoietic unity of second order. This has actually happened on earth with the evolution of the multicellular pattern of organization. When this occurs, the component (living) autopoietic systems become necessarily subordinated, in the way they realize their autopoiesis, to the maintenance of the autopoiesis of the higher order autopoietic unity which, through their coupling, they define topologically in the physical space. If the higher order autopoietic system undergoes self-reproduction (through the self-reproduction of one of its component autopoietic unities or otherwise), an evolutionary process begins in which the evolution of the manner of realization of the component autopoietic systems is necessarily subordinated to the evolution of the manner of realization of the composite unity. Furthermore, it is to be expected that if the proper contingencies are given, higher order autopoietic unities will be formed through selection. In fact, if coupling arises as a form of satisfying autopoiesis, a second order unity formed from previous autopoietic systems will be more stable, the more stable the coupling is. However, the most stable condition for coupling appears if the unity organization is precisely geared to maintain this organization, this is, if the unity becomes autopoietic. There is then an ever present selective pressure for the constitution of higher order autopoietic systems from the coupling of lower order autopoietic unities which on earth is apparent in the occurrence of multicellular systems, if not in that of the eucariotic cell itself. It seems that the only limit to the process of constitution of autopoietic unities of higher order is that imposed by the circumstances under which a unity can be specified in a given space.

CHAPTER V

PRESENCE OF AUTOPOIESIS

Autopoiesis in the physical space is necessary and sufficient to characterize a system as a living system. Reproduction and evolution as they occur in the known living systems, and all the phenomena derived from them, arise as secondary processes subordinated to their existence and operation as autopoietic unities. Hence, the biological phenomenology is the phenomenology of autopoietic systems in the physical space, and a phenomenon is a biological phenomenon only to the extent that it depends in one way or another on the autopoiesis of one or more physical autopoietic unities.

1. BIOLOGICAL IMPLICATIONS

A living system is a living system because it is an autopoietic system in the physical space, and it is a unity in the physical space because it is defined as a unity in that space by and through its autopoiesis. Accordingly, any structural transformation that a living system may undergo maintaining its identity must take place in a manner determined by and subordinated to its defining autopoiesis; hence, in a living system loss of autopoiesis is disintegration as a unity and loss of identity, that is, death.

(i) The physical space is defined by components that can be determined by operations that characterize them in terms of properties such as masses, forces, accelerations, distances, fields, etc. Furthermore, such properties themselves are defined by the interactions of the components that they characterize. In the physical space two kinds of phenomenologies can take place according to the way the components participate in their generation, namely, statical and mechanical (machine like). The statical phenomenology is a phenomenology of relations between properties of components; the mechanical phenomenology is a phenomenology of relations between processes realized through the properties of components. What about the biological phenomenology, that is, what about the phenomenology of autopoietic systems, which, as such, takes place in the physical space? Since a living system is defined as a system by the concatenation of processes of production of components that generate the processes that produce them and constitute the system as a unity in the physical space, biological phenomena are necessarily phenomena of relations between processes which satisfy the autopoiesis of the participant living systems. Accordingly, under no circumstances is a biological phenomenon defined by the properties of its component elements, but it is always defined and constituted by a concatenation of processes in relations subordinated to the autopoiesis of at least one living system. Thus, the accidental collision of two running animals, as a bodily encounter of living systems, is not a biological phenomenon (even though it may have biological consequences), but the bodily contact of two animals in courtship is. Strictly, then, although biological and statical phenomena are physical phenomena because they are realized through the properties of their physical components, they differ because statical phenomena are phenomena of relations between properties of components (as previously defined), while biological phenomena are phenomena of relations between processes. Therefore, biological phenomena as phenomena of relations between processes are a subclass of the mechanical phenomena which constitute them, and are defined through the participation of these processes in the realization of at least one autopoietic system. The phenomenology of living systems, then, is the mechanical phenomenology of physical autopoietic machines.

(ii) As the mechanical phenomenology of physical autopoietic machines, the biological phenomenology is perfectly defined, and, hence, amenable to theoretical treatment through the theory of autopoiesis. It follows that such a theory as a formal theory will be a theory of the concatenation of processes of production that constitute autopoietic systems, and not a theory of properties of components of living systems. It also follows that a theoretical biology would be possible as a theory of the biological phenomenology, and not as the application of physical or chemical notions, which pertain to a different phenomenological domain, to the analysis of the biological phenomena. In fact, it should be apparent now that any attempt to explain a biological phenomenon in statical or non-autopoietic mechanical terms would be an attempt to reformulate it in terms of relations between properties of components, or relations between processes which do not involve an autopoietic unity in the physical space, and would fail to reformulate it. Since a biological phenomenon takes place through the operation of components, it is always possible to abstract from it component processes that can be adequately described in statical or non-autopoietic mechanical terms, because, as abstracted processes, they in fact correspond to statical or allopoietic mechanical phenomena. In such a case, any connection between the statical or non-autopoietic mechanical processes and the biological phenomenon from which the observer abstracts them, is provided by the observer who considers both simultaneously; the biological phenomenon, however, is not and cannot be captured by these explanations which, necessarily, remain a reformulation of a phenomenon in a non-autopoietical phenomenological domain. A biological explanation must be a reformulation in terms of processes subordinated to autopoiesis, that is, a reformulation in the biological phenomenological domain.

(iii) An adequate theory of the biological phenomena should permit the analysis of the dynamics of the concrete components of a system in order to determine whether or not they participate in processes that integrate a biological phenomenon. In fact, no matter how much we think we understand biological problems today, it is apparent that without an adequate theory of autopoiesis it will not be possible to answer questions such as: 'Given a dynamic system, what relations should I observe between its concrete components to determine whether or not they participate in processes that make it a living system?'; or, 'Given a set of components with well-defined properties, in what processes of production can they participate so that the components can be concatenated to form an autopoietic system?' The answers to these questions are essential if one wants to solve the problem of the origin of living systems on earth. The same questions must be answered if one wants to design a living system. In particular, it should be possible to determine from biological theoretical considerations which relations should be satisfied by any set of components if these are to participate in processes that constitute an autopoietic unity. Whether one may or may not want to make an autopoietic system is, of course, a problem that pertains to the ethical domain. However, if our characterization of living systems is adequate it is apparent that they could be made at will. What remains to be seen is whether such a system has already been made by man, although unwittingly, and with what consequences.

(iv) The characterization of living systems as physical autopoietic systems must be understood as having universal value, that is, autopoiesis in the physical space must be viewed as defining living systems anywhere in the universe, however different they may otherwise be from terrestrial ones. This is not to be considered as a limitation of our imagination, nor as a denial that there might exist still unimagined complex systems. It is a statement about the nature of the biological phenomenology: the biological phenomenology is not less and not more than the phenomenology of autopoietic systems in the physical space.

2. EPISTEMOLOGICAL IMPLICATIONS

(i) The basic epistemological question in the domain of the biological problems is that which refers to the validity of the statements made about biological systems. It is presently obvious that scientific statements made about the universe acquire their validity through their operative effectiveness in their application in the domain where they pretend validity. Yet any observation, even that one which permits us to recognize the operational validity of a scientific statement, implies an epistemology, a body of conceptual explicit or implicit notions that determines the perspective of the observations and, hence, what can and what cannot be observed, what can and what cannot be validated by its operative effectiveness, what can and what cannot be explained by a given body of theoretical concepts. This has been a fundamental problem in the conceptual and experimental handling of the biological phenomena, as it is apparent in the history of biology, which reveals a continuous search for the definition of the biological phenomenology in a manner such that would permit its complete explanation through well-defined notions, and, accordingly, its complete validation in the observational domain. In this respect, evolutionary and genetic notions have been the most successful. Yet these notions alone are insufficient because, although they provide a mechanism for historical change, they do not adequately define the domain of the biological phenomenology. In fact, evolutionary and genetic notions (by emphasizing generational change) treat the species as the source of all biological order, showing that the species evolves while the individuals are transient components whose organization is subordinated to its historical phenomenology. However, since the species is, concretely at any moment, a collection of individuals capable in principle of interbreeding, it turns out that what would define the organization of individuals is either an abstraction, or something that requires the existence of well-defined individuals to begin with. Where does the organization of the individual come from? Which is the mechanism for its determination? This difficulty cannot be solved on purely evolutionary and genetic arguments, since it is apparent (even for evolutionists and geneticists) that any attempt to overcome it by resorting to other notions of comprehensive nature, is doomed to failure if it does not provide us with a mechanism to account for the phenomenology of the individual. Such is the case when some sort of preformism is introduced by applying informational notions at the molecular level (nucleic acids or proteins); or when organismic notions are used that emphasize the unitary character of living systems but do not provide a mechanism for the definition of the individual. These notions fail because they imply the validity of the same notion that they want to explain.

As is apparent from all that has been said, the key to the understanding of the biological phenomenology is the understanding of the organization of the individual. We have shown this organization to be the autopoietic organization. Furthermore, we have shown that this organization and its origin are fully explainable with purely mechanistic notions which are valid for any mechanistic phenomenon in any space, and that once the autopoietic organization is established it determines an independent phenomenological subdomain of the mechanistic phenomenology, the domain of the biological phenomena. As a result, the biological domain is fully defined and self-contained, no additional notions are necessary, and any adequate biological explanation has the same epistemological validity that any mechanistic explanation of any mechanistic phenomenon in the physical space has.

(ii) A phenomenological domain is defined by the properties of the unity or unities that constitute it, either singly or collectively through their transformations or interactions. Thus, whenever a unity is defined, or a class or classes of unities are established which can undergo transformations or interactions, a phenomenological domain is defined. Two phenomenological domains intersect only to the extent that they have common generative unities, that is, only to the extent that the unities that specify them interact; otherwise they are completely independent and, obviously, they cannot generate each other without transgressing the domains of relations of their respective specifications. Conversely, one phenomenological domain can generate unities that define a different phenomenological domain, but such a domain is specified by the properties of the new different unities, not by the phenomenology that generates them. If this were not the case the new unities would not be in fact different unities, but they would be unities of the same class of units that generate the parental phenomenological domain, and they would generate a phenomenological domain identical to it. Autopoietic systems do generate different phenomenological domains by generating unities whose properties are different from the properties of the unities that generate them. These new phenomenological domains are subordinated to the phenomenology of the autopoietic unities because they depend on these for their actual realization, but they are not determined by them; they are only determined by the properties of their originating unities regardless of how these were originated. One phenomenological domain cannot be explained by relations which are valid for another domain; this is a general case which applies also to the different phenomenological domains generated through the operation of autopoietic systems. Accordingly, as an autopoietic system cannot be explained through statical or non-autopoietic mechanical relations in the space in which it exists, but it must be explained through autopoietic mechanical relations in the mechanical domain, the phenomena generated through interactions of autopoietic unities must be explained in the domain of interactions of the autopoietic unities through the relations that define that domain.

(iii) The development of the Darwinian notion of evolution with its emphasis on the species, natural selection and fitness, had an impact in human affairs that went beyond the explanation of diversity and its origin in living systems. It had sociological significance because it seemed to offer an explanation of the social phenomenology in a competitive society, as well as a scientific justification for the subordination of the destiny of the individuals to the transcendental values supposedly embodied in notions such as mankind, the state, or society. In fact, the social history of man shows a continuous search for values that explain or justify human existence, as well as a continuous use of transcendental notions to justify social discrimination, slavery, economical subordination and political submission of the individuals, isolated or collectively, to the design or whim of those who pretend to represent the values contained in those notions. For a society based on economic discrimination, competitive ideas of power and subordination of the citizen to the state, the notions of evolution, natural selection and fitness (with their emphasis on the species as the perduring historical entity maintained through the dispensability of transient individuals) seemed to provide a biological (scientific) justification for its economic and social structure. It is true on biological grounds that what evolves is mankind as the species *Homo sapiens*. It is true on biological grounds that competition participates in the specification of evolutionary change even in man. It is true that under the laws of natural selection the individuals most apt in the features which are favorably selected survive, or have reproductive advantages over the others, and that those which do not survive or are less successful in the reproductive sense do not contribute or contribute less to the historical destiny of the species. Thus, from the Darwinian perspective it seemed that the role of the individual was to contribute to the perpetuation of the species, and that all that one had to do for the well-being of mankind was to let the natural phenomena follow their course. Science, biology, appeared to justify the notion 'anything for the benefit of mankind', whatever the intention or purpose of whoever uttered it first. We have shown, however, that these arguments are not valid to justify the subordination of the individual to the species, because the biological phenomenology is determined by the phenomenology of the individuals, and without individuals there is no biological phenomenology whatsoever. The organization of the individual is autopoietic and upon this fact rests all its significance: it becomes defined through its existing, and its existing is autopoietic. Thus, biology cannot be used anymore to justify the dispensability of the individuals for the benefit of the species, society or mankind under the pretense that its role is to perpetuate them. Biologically the individuals are not dispensable.

(iv) Biological phenomena depend upon the autopoiesis of the individuals involved; thus, there are biological systems that arise from the coupling of autopoietic unities, some of which may even constitute autopoietic systems of higher order. What about human societies, are they, as systems of coupled human beings, also biological systems? Or, in other words, to what extent do the relations which characterize a human society as a system constitutively depend on the autopoiesis of the individuals which integrate it? If human societies are biological systems the dynamics of a human society would be determined through the autopoiesis of its components. If human societies are not biological systems, the social dynamics would depend on laws and relations which are independent of the autopoiesis of the individuals which integrate them. The answer to this question is not trivial and requires considerations which in addition to their biological significance have ethical and political implications. This is obviously the case, because such an answer requires the characterization of the relations which define a society as a unity (a system), and whatever we may say biologically will apply in the domain of human interactions directly, either by use or abuse, as we saw it happen with evolutionary notions. In fact no position or view that has any relevance in the domain of human relations can be deemed free from ethical and political implications nor can a scientist consider himself alien to these implications. This responsibility we are ready to take, yet since we – Maturana and Varela – do not fully agree on an answer to the question posed by the biological character of human societies from the vantage point of this characterization of the biological organization, we have decided to postpone this discussion.

3. COGNITIVE IMPLICATIONS

The domain of interactions of an autopoietic unity is the domain of all the deformations that it may undergo without loss of autopoiesis. Such a domain is determined for each unity by the particular mode through which its autopoiesis is realized in the space of its components, that is, by its structure. It follows that the domain of interactions of an autopoietic unity is necessarily bounded, and that autopoietic unities with different structures have different domains of interactions. Furthermore, an observer can consider the way in which an autopoietic system compensates its deformations as a description of the deforming agent that he sees acting upon it, and the deformation suffered by the system as a representation of the deforming agent. However, since the domain of interactions of an autopoietic system is bounded, an observer of an autopoietic system can describe entities external to it (by interacting with them) which it cannot describe because it cannot interact with them or it cannot compensate the deformations which these cause in it. The domain of all the interactions in which an autopoietic system can enter without loss of identity is its cognitive domain; or, in other words, the cognitive domain of an autopoietic system is the domain of all the descriptions which it can possibly make. Accordingly, for any autopoietic system its particular mode of autopoiesis determines its cognitive domain and hence its behavioral diversity, and it follows that the cognitive domain of an autopoietic system changes along its ontogeny only to the extent that its mode of autopoiesis changes.

We shall not explore in this book all the implications that the proper characterization of the biological phenomenology has within the domain of cognition, but we shall make four remarks in order to show the dependence of this domain upon the autopoietic organization of the individual.

(i) For any autopoietic system its cognitive domain is necessarily relative to the particular way in which its autopoiesis is realized. Also, if knowledge is descriptive conduct, it is relative to the cognitive domain of the knower. Therefore, if the way in which the autopoiesis of an organism is realized changes during its ontogeny, the actual knowledge of the organism (its conduct repertoire) also changes; knowledge, then, is necessarily always a reflection of ontogeny of the knower because ontogeny as a process of continuous structural change without loss of autopoiesis is a process of continuous specification of the behavioral capacity of the organism, and, hence, of its actual domain of interactions. Intrinsically, then, no absolute knowledge is possible, and the validation of all possible relative knowledge is attained through successful autopoiesis.

(ii) Autopoietic systems may interact with each other under conditions that result in behavioral coupling. In this coupling, the autopoietic conduct of an organism A becomes a source of deformation for an organism B, and the compensatory behavior of organism B acts, in turn, as a source of deformation of organism A, whose compensatory behavior acts again as a source of deformation of B, and so on recursively until the coupling is interrupted. In this manner, a chain of interlocked interactions develops such that, although in each interaction the conduct of each organism is constitutively independent in its generation of the conduct of the other, because it is internally determined by the structure of the behaving organism only, it is for the other organism, while the chain lasts, a source of compensable deformations which can be described as meaningful in the context of the coupled behavior. These are communicative interactions. In other words, if the interacting organisms as dynamic systems have continuously changing structures, and if they reciprocally select in each other their respective paths of ontogenic structural changes through their interactions without loss of autopoiesis, then they generate, as a recursive or expanding domain of communicative interactions, interlocked ontogenies that together constitute a domain of mutually triggering consensual conducts that becomes specified during its generation. Such a consensual domain of communicative interactions in which the behaviorally coupled organisms orient each other with modes of behavior whose internal determination has become specified during their coupled ontogenies, is a linguistic domain. In such a consensual domain of interactions the conduct of each organism may be treated by an observer as constituting a connotative description of the conduct of the other, or, in his domain of description as an observer, as a consensual denotation of it. Communicative and linguistic interactions are intrinsically not informative; organism A does not and cannot determine the conduct of organism B because due to the nature of the autopoietic organization itself every change that an organism undergoes is necessarily and unavoidably determined by its own organization. A linguistic domain, then, as a consensual domain that arises from the coupling of the ontogenies of otherwise independent autopoietic systems, is intrinsically non-informative, even though an observer, by neglecting the internal determination of the autopoietic systems which generate it, may describe it as if it were so. Phenomenologically the linguistic domain and the domain of autopoiesis are different domains, and although one generates the elements of the other, they do not intersect.

(iii) An autopoietic system capable of interacting with its own states (as an organism with a nervous system can do), and capable of developing with others a linguistic consensual domain, can treat its own linguistic states as a source of deformations and thus interact linguistically in a closed linguistic domain. Such a system has two remarkable properties:

1. Through recursive interactions with its linguistically generated states it can treat some of these states as objects of further interactions, giving rise to a metadomain of consensual distinctions that appears to an observer as a domain of interactions with representations of interactions. When this happens the system operates as an observer. The domain of such recursive interactions is, in principle, infinite because once the system has attained the mechanism for doing so there is no moment in which it will not be in the position of recursively interacting with its own states, unless autopoiesis is lost. Whether an autopoietic system with this capacity does in fact generate an endless series of different states during its ontogeny depends, obviously, on whether its history of linguistic interactions in the metadomain of descriptions has significance for the circumstantial realization of the autopoiesis of the interacting organisms.

2. A living system capable of being an observer can interact with those of its own descriptive states which are linguistic descriptions of itself. By doing so it generates the domain of self-linguistic descriptions within which it is an observer of itself as an observer, a process which can be necessarily repeated in an endless manner. We call this domain the domain of self-observation and we consider that self-conscious behavior is self-observing behavior, that is, behavior within the domain of self-observation. The observer as an observer necessarily always remains in a descriptive domain, that is, in a relative cognitive domain. No description of an absolute reality is possible. Such a description would require an interaction with the absolute to be described, but the representation which would arise from such an interaction would necessarily be determined by the autopoietic organization of the observer, not by the deforming agent; hence, the cognitive reality that it would generate would unavoidably be relative to the knower.

In every explanation, be this an actual concrete reproduction, a formal representation or a purely rational description, the reformulation of the phenomenon to be explained resorts to the same notions (identity, exclusion, succession, etc.). There is, then, a universal logic, valid for all phenomenological domains, that refers to the relations possible between the unities that generate these domains, and not to the particular properties of the generating unities. We have applied this logic (it could not have been otherwise) in this book, and the validity of our arguments, as the validity of any rational argument or concrete phenomenological realization, rests on its validity. Furthermore, we have in principle shown through its application that the phenomenology of autopoietic systems generates observers, and through them the phenomenology of description within which this logic is also valid. For epistemological reasons, in order to say all that we have said about living systems, we had to assume a space (the physical space) within which the phenomenology of autopoiesis of living systems takes place. To the extent that we have been successful (free from logical and experiential contradictions), we can conclude that such a space is ontologically a space within which the logic that we have applied in our description is intrinsically valid. If this were not the case we could not have done what we have done in terms of characterizing living systems, or of showing how these may generate systems capable of their own description. We cannot characterize this space in absolute terms. In linguistic interactions, all that we can do is to describe through linguistic behavior and construct further descriptions based on these descriptions which always remain in the same domain of operations defined in relation to the operating system.

A prediction is a statement of a case within a relational matrix; it is a cognitive statement, and as such it takes place within a descriptive domain. Thus, unless mistakes are made, if all the relations that define the particular matrix within which the prediction is made are properly taken, the prediction is valid. Errors of interpretations may arise only by mis-application, that is, by pretending that the observer makes a prediction in one matrix when he is making it in another. In particular, predictions in the physical space are possible, because a description, as an actual behavior, exists in a matrix of interactions which (by constitution) has a logical matrix necessarily isomorphic with the substratum matrix within which it takes place, not because we have an absolute knowledge of the universe. These cognitive relations are valid for the possible cognitive phenomenology generated by any closed system. Living systems are an existential proof; they exist only to the extent that they can exist. The fantasy of our imagination cannot deny this. Living systems are concatenations of processes in a mechanistic domain; fantasies are concatenations of descriptions in a linguistic domain. In the first case, the concatenated unities are processes; in the second case, they are modes of linguistic behavior.

Autopoiesis solves the problem of the biological phenomenology in general by defining it. New problems arise, and old ones appear in a different perspective; in particular, those which refer to the origin of living systems on earth (eobiogenesis and neobiogenesis), and those which refer to the particular organization through which recursive descriptive interactions take place in animals (the nervous system). Autopoietic systems define the world in which they can exist in relation to their autopoiesis, and some interact recursively with this world through their descriptions, it being impossible for them to step out of this relative descriptive domain through descriptions. This demands an entirely new cognitive outlook: there is a space in which different phenomenologies can take place; one of these is autopoiesis; autopoiesis generates a phenomenological domain, this is cognition.

APPENDIX

THE NERVOUS SYSTEM

The phenomenology of an organism as a unity is the phenomenology of its autopoiesis. The changes that an organism undergoes while maintaining its autopoiesis constitute its conduct. The conduct of an organism is revealed to an observer by the changes that it causes in the ambience (including the observer) in which it exists. Accordingly, the conduct which an observer beholds in any organism, however complex it may seem, is always an expression of the autopoiesis of the observed organism, and as such, it always arises through a phenomenology that takes places in the present because history is not a causal component in the mechanism of autopoiesis (see Chapter IV). Yet it appears to us as subjects of self-observation and as observers of the conduct of other organisms that past experiences determine our and their conduct in the present as if, embodied in modifications of the nervous system, they were causal components in the mechanism which generates behavior. It appears, therefore, as if the operation of the organism as a state-determined system in which time is not a component were determined by temporal phenomena, and we speak of learning, memory and recall as embodiments of the past. We consider that this contradiction arises from not distinguishing what pertains to the phenomenology of the autopoiesis from what pertains to the domain of interactions of the organism as a unity, and, thus, from an inadequate evaluation of the coupling of the structure of the nervous system to the ontogeny of the organism. Accordingly, our purpose in this Appendix about the nervous system is to consider its organization as a neuronal network and to evaluate this coupling in which past and present arise as new dimensions from the recursive interactions of the organism with its own states.

A. THE NERVOUS SYSTEM AS A SYSTEM

The nervous system is a network of interacting neurons coupled in three ways to the organism of which it is a component:

(i) The organism, including the nervous system, provides the physical and biochemical environment for the autopoiesis of the neurons as well as for all other cells, and, hence, is a possible source of physical and biochemical perturbations which may alter the properties of the neurons and thus lead to (ii) or (iii).

(ii) There are states of the organism (physical and biochemical) which change the state of activity of the nervous system as a whole by acting upon the receptor surfaces of some of its component neurons, and thus lead to (iii).

(iii) There are states of the nervous system which change the state of the organism (physical or biochemical) and lead recursively to (i) and (ii).

Through this coupling the nervous system participates in the generation of the autopoietic relations which define the organism which it integrates, and, accordingly, its structure is subordinated to this participation.

1. *The Neuron*

Neurons determine their own boundaries through their autopoiesis; therefore they are the anatomical units of the nervous system. There are many classes of neurons that can be distinguished by their shapes, but all of them, regardless of the morphological class to which they belong, have branches which put them in direct or indirect operational relations with other otherwise separated neurons. Functionally, that is, viewed as an allopoietic component of the nervous system, a neuron has a collector surface, a conducting element, and an effector surface, whose relative positions, shapes and extensions are different in different classes of neurons. The collector surface is that part of the surface of a neuron where it receives afferent influences (synaptic or not) from the effector surfaces of other neurons or its own. The effector surface of a neuron is that part of its surface which either directly (by means of synaptic contacts) or indirectly (through its synaptic or nonsynaptic action on other kinds of cells) affects the collector surface of other neurons or its own. Depending on its kind, a neuron may have its collector and effector surfaces completely or partly separated by a conducting element (absence or presence of presynaptic inhibition), or it may have both collector and effector surfaces completely interspaced, with no conducting element between them (amacrine cells). The interactions between collector and effector surfaces may be excitatory or inhibitory according to the kinds of neurons involved. Excitatory afferent influences cause a change in the state of activity of the collector surface of the receiving neuron which may lead to a change in the state of activity of its effector surface, while the inhibitory influences impinging on it shunt off the effect of the afferent influences on its receptor surface so that this effect does not at all reach its effector surface, or reaches this surface with reduced effectiveness.

Operationally the state of activity of a neuron, characterized by the state of activity of its effector surface, is determined by both its internal structure (membrane properties, relative thickness of branches, and in general all structural relations which determine its possible states) and the afferent influences impinging on its receptor surface. Conversely, the effectiveness of a neuron in changing the state of activity of other neurons depends both on the internal structure of these, and on the relative effectiveness of its action on their receptor surfaces with respect to the other afferent influences that these neurons receive. This is so because excitatory and inhibitory influences do not add linearly in the determination of the state of activity of a neuron, but rather have effects which depend on the relative position of their points of action with respect to each other and with respect to the effector surface of the receiving cell. Furthermore, the internal structure of a neuron changes along its life history, both as a result of its autonomous genetic determinations and as a result of the circumstances of its operations during the ontogeny of the organism. Thus, neurons are not static entities whose properties remain invariant. On the contrary, they change. This has three general consequences:

(i) There are many configurations of afferent (input) influences on the receptor surface of a neuron which produce the same configuration of efferent (output) activity at its effector surface.

(ii) Changes in the internal structure of a neuron (regardless of whether they arise selected by the autonomous transformation of the cell, or by its history of interactions in the neuronal network) by changing the domain of states of activity that the neuron can adopt, change its domain of input-output relations, that is, change its transfer function.

(iii) No single cell or class of cells can alone determine the properties of the neural network which it integrates.

Generally then, the structure of a neuron and its role in the neuronal network which it integrates does not stay invariant, but changes along its ontogeny in a manner subordinated to the ontogeny of the organism which is both a result and a source of the changes that the neuronal network and the organism undergo.

2. *Organization*: *The Nervous System As a Closed System*

From the descriptive point of view it is possible to say that the properties of the neurons, their internal structure, shape and relative position, determine the connectivity of the nervous system and constitute it as a dynamic network of neuronal interactions. This connectivity, that is, the anatomical and operational relations which hold between the neurons which constitute the nervous system as a network of lateral, parallel, sequential and recursive inhibitory and excitatory interactions, determines its domain of possible dynamic states. Since the properties of the neurons change along the ontogeny of the organism, both due to their internal determination and as a result of their interactions as components of the nervous system, the connectivity of the nervous system changes along the ontogeny of the organism in a manner recursively selected during this ontogeny. Furthermore, since the ontogeny of the organism is the history of its autopoiesis, the connectivity of the nervous system, through the neurons which constitute it, is dynamically subordinated to the autopoiesis of the organism which it integrates.

Operationally, the nervous system is a closed network of interacting neurons such that a change of activity in a neuron always leads to a change of activity in other neurons, either directly through synaptic action, or indirectly through the participation of some physical or chemical intervening element. Therefore, the organization of the nervous system as a finite neuronal network is defined by relations of closeness in the neuronal interactions generated in the network. Sensory and effector neurons, as they would be described by an observer who beholds an organism in an environment, are not an exception to this because all sensory activity in an organism leads to activity in its effector surfaces, and all effector activity in it leads to changes in its sensory surfaces. That at this point an observer should see environmental elements intervening between the effector and the sensory surfaces of the organism, is irrelevant because the nervous system is defined as a network of neuronal interactions by the interactions of its component neurons regardless of intervening elements. Therefore, as long as the neural network closes on itself, its phenomenology is the phenomenology of a closed system in which neuronal activity always leads to neuronal activity. This is so even though the ambience can perturb the nervous system and change its status by coupling to it as an independent agent at any neuronal receptor surface. The changes that the nervous system can undergo without disintegration (loss of defining relations as a closed neuronal network) as a result of these or any other perturbation are fully specified by its connectivity, and the perturbing agent only constitutes a historical determinant for the concurrence of these changes. As a closed neuronal network the nervous system has no input or output, and there is no intrinsic feature in its organization which would allow it to discriminate through the dynamics of its changes of state between possible internal or external causes for these changes of state. This has two fundamental consequences:

(i) The phenomenology of the changes of state of the nervous system is exclusively the phenomenology of the changes of state of a closed neuronal network; that is, for the nervous system as a neuronal network there is no inside or outside.

(ii) The distinction between internal and external causes in the origin of the changes of state of the nervous system can only be made by an observer that beholds the organism (the nervous system) as a unity, and defines its inside and outside by specifying its boundaries.

It follows that it is only with respect to the domain of interactions of the organism as a unity that the changes of state of the nervous system may have an internal or an external origin, and, hence, that the history of the causes of the changes of state of the nervous system lies in a different phenomenological domain than the changes of state themselves.

3. *Change*

Any change in the structure of the nervous system arises from a change in the properties of its component neurons. What change in fact takes place, whether morphological or biochemical or both, is irrelevant for the present discussion. The significant point is that these changes arise in the coupling of the nervous system and the organism through their homeostatic operation subordinated to the autopoiesis of the latter. Some of the changes directly affect the operation of the nervous system because they take place through its working as a closed network; others affect it indirectly because they take place through the biochemical or genetic coupling of the neurons to the organism and change the properties of the neurons in a manner unrelated to the actual working of the network. The results are twofold: on the one hand, all changes lead to the same thing, that is, changes in the domain of possible states of the nervous system; on the other hand the nervous system is coupled to the organism both in its domain of interactions and in its domain of internal transformations.

4. *Architecture*

The connectivity of the nervous system is determined by the shapes of its component neurons. Accordingly, every nervous system has a definite architecture determined by the kinds and the numbers of the neurons which compose it; therefore, members of the same species have nervous systems with similar architectures to the extent that they have similar kinds and numbers of neurons. Conversely, members of different species have nervous systems with different architectures according to their specific differences in neuronal composition. Therefore, the closed organization of the nervous system is realized in different species in different manners that have been determined through evolution; in all cases, however, the following conditions are satisfied:

(i) Since, due to its constitution as a network of lateral, parallel, sequential and recursive interactions, the nervous system closes on itself at all levels, the mutilations that it may suffer generally leave a closed neuronal network with a changed architecture. Accordingly, the organization of the nervous system is essentially invariant under mutilations, while its domain of possible states, which depends on its structure, and, hence, on its architecture, is not. Yet, due to its closed organization, whatever is left of the neural network after a partial ablation necessarily operates as a different whole with different properties than the original, but not as a system to which some of its properties have been selectively subtracted.

(ii) There is intrinsically no possibility of operational localization in the nervous system in the sense that no part of it can be deemed responsible for its operation as a closed network, or for the properties which an observer can detect in its operation as a unity. However, since every nervous system has a definite architecture, every localized lesion in it necessarily produces a specific disconnection between its parts and, hence, a specific change in its domain of possible states.

(iii) The architecture of the nervous system is not static, but it becomes specified along the ontogeny of the organism to which it belongs, and its determination, although under genetic control, is bound to the morphogenesis of the whole organism. This has two implications: (a) the variability in the architecture of the nervous system of the members of a species is determined by individual differences in genetic constitution and ontogeny; (b) the range of permissible individual variations (compatible with the autopoiesis) is determined by the circumstances in which the autopoiesis of the organism is realized.

(iv) The architecture of the nervous system and the morphology of the organism as a whole define the domain in which the ambience can possibly couple on the organism as a source of its deformations. Thus, as long as the architecture of the nervous system and the morphology of the organism remain invariant, or as long as there are aspects of them which remain unchanged, there is the possibility of recurrent perturbations as recurrent configurations of the ambience which couple in the same way on the nervous system and the organism.

5. *Referential States*

There are states of the nervous system which, as referential states, define subdomains of the possible states that the nervous system (and the organism) can adopt under perturbations as matrices of possible internal relations. As a result when the nervous system is in different referential states it compensates the same perturbations (characterized as configuration of the ambience) following different characteristic modes of change. Emotions, sleep, wakefulness, are referential states. In the dynamics of the nervous system, referential states are defined as are all other states of the nervous system, that is, by relations of neuronal activity, and as such are generated by change of neuronal activity and generate changes of neuronal activity. What is peculiar to them is that they constitute states on which other states can be inserted as substates in the process of generating the autopoiesis of the organisms. Therefore, their distinction lies in the domain of observation because for the nervous system they are part of their dynamic of state to state operations, and in the domain of observation they constitute independent phenomenological dimensions.

B. CONSEQUENCES

1. *Historical Coupling*

Due to its coupling with the organism the nervous system necessarily participates in the generation of the relations which constitute the organism as an autopoietic unity. Also due to this coupling the structure of the nervous system is necessarily continuously determined and realized through the generation of neuronal relations internally defined with respect to the nervous system itself. As a consequence, the nervous system necessarily operates as an homeostatic system that maintains invariant the relations which define its participation in the autopoiesis of the organism, and does so by generating neuronal relations which are historically determined along the ontogeny of the organism through its participation in this ontogeny. This has the following implications:

(i) The changes that the nervous system undergoes as an homeostatic system while compensating the deformations that it suffers as a result of the interactions of the organism (itself an homeostatic system), cannot be localized to any singular point in the nervous system, but must be distributed along it in a non-random manner because any localized change is itself a source of additional deformations that must be compensated with further changes. This process is potentially endless. As a result, the operation of the nervous system as a component of the organism is a continuous generation of significant neuronal relations, and all the transformations that it may undergo as a closed neuronal network are subordinated to this. If as a result of a perturbation the nervous system fails in the generation of the significant neuronal relations for its participation in the autopoiesis of the organism, the organism disintegrates.

(ii) Although the organism and nervous system are closed atemporal systems, the fact that the structure of the nervous system is determined through its participation in the ontogeny of the organism makes this structure a function of the circumstances which determine this ontogeny, that is, of the history of interactions of the organism as well as of its genetic determination. Therefore, the domain of the possible states that the nervous system can adopt as an atemporal system is at any moment a function of this history of interactions and implies it. The result is the coupling of two constitutively different phenomenologies, the phenomenology of the nervous system (and the organism) as a closed homeostatic system, and the phenomenology of the ambience (including the organism and the nervous system) as an open non-homeostatic system which thus braid together in a manner such that the domain of the possible states of the nervous system continuously becomes commensurate with the domain of the possible states of the ambience. Furthermore, since all states of the nervous system are internal states, and the nervous system cannot make a distinction in its process of transformation between its internally and externally generated changes, the nervous system is bound to couple its history of transformations to the history of its internally determined changes of state as much as to the history of its externally determined changes of state. Thus the transformations that the nervous system undergoes during its operation are a constitutive part of its ambience.

(iii) The historical coupling of the nervous system to the transformations of its ambience, however, is apparent only in the domain of observation, not in the domain of operation of the nervous system which remains a closed homeostatic system in which all states are equivalent to the extent that they all lead to the generation of the relations which define its participation in the autopoiesis of the organism. The observer can see that a given change in the structure of the nervous system arises as a result of a given interaction of the organism, and he can consider this change as a representation of the circumstances of the interaction. The representation, however, as a phenomenon exists only in the domain of observation and has a validity that applies only in the domain generated by the observer as he maps the environment on the behaviors of the organism by treating it as an allopoietic system. The referred change in structure of the nervous system constitutes a change in the domain of its possible states under conditions in which the representation of the causing circumstances do not enter as a component.

2. *Learning as a Phenomenon*

If the connectivity structure of the nervous system changes as a result of some interactions of the organism, the domain of the possible states which it (and the organism) can henceforth adopt, changes; as a consequence, when the same or similar conditions of interaction recur, the dynamic states of the nervous system and, therefore, the way the organisms attains autopoiesis are necessarily different from what they would have otherwise been. Yet, that the conduct of the organism under the recurrent (or new) conditions of interaction should be autopoietic and, hence, appear adaptive to an observer, is a necessary outcome of the continuous homeostatic operation of both the nervous system and the organism. Since this homeostatic operation continuously subordinates the nervous system and the organism to the latter's autopoiesis in an internally determined manner, no change of connectivity in the nervous system can participate in the generation of behavior as a representation of the past interactions of the organism: representations belong to the domain of descriptions. The change in the domain of the possible states that the nervous system can adopt, which takes place along the ontogeny of the organism as a result of its interactions, constitutes learning. Thus, learning as a phenomenon of transformation of the nervous system associated to a behavioral change that takes place under maintained autopoiesis, occurs due to the continuous dynamic coupling of the state-determined phenomenology of the nervous system and the state-determined phenomenology of the ambient. The notions of acquisition of representations of the environment or of acquisition of information about the environment in relation to learning, do not represent any aspect of the operation of the nervous system. The same applies to notions such as memory and recall, which are descriptions made by the observer of phenomena that take place in his domain of observation, and not in the domain of operation of the nervous system, and, hence, have validity only in the domain of descriptions, where they are defined as causal components.

3. *Time as a Dimension*

Any mode of behavioral distinction between otherwise equivalent interactions, in a domain that has to do with the states of the organism and not with the ambience features which define the interaction, gives rise to a referential dimension as a mode of conduct. This is the case with time. It is sufficient that as a result of an interaction (defined by an ambience configuration) the nervous system should be modified with respect to the specific referential state (emotion of assuredness, for example) which the recurrence of the interaction (regardless of its nature) may generate for otherwise equivalent interactions to cause conducts which distinguish them in a dimension associated with their sequence, and, thus, give rise to a mode of behavior which constitutes the definition and characterization of this dimension. Therefore, sequence as a dimension is defined in the domain of interactions of the organism, not in the operation of the nervous system as a closed neuronal network. Similarly, the behavioral distinction by the observer of sequential states in his recurrent states of nervous activity, as he recursively interacts with them, constitutes the generation of time as a dimension of the descriptive domain. Accordingly, time is a dimension in the domain of descriptions, not a feature of the ambience.

C. IMPLICATIONS

Since history as a phenomenon is accessible to the observer only in the domain of descriptions, it is only in this domain that history may participate in the generation of the observer's behavior. This, in fact, takes place. Descriptions as linguistic behavior constitute a source of deformations of the nervous system and, hence, part of its ambience. Accordingly, the phenomenology of transformation of the nervous system discussed above also applies to the interactions of the organism in the domain of descriptions, and the structure of the nervous system is also a function of the history of interactions of the organism in this domain. The implications are obvious. The operation of the nervous system makes no distinction between its different sources of deformation, and, accordingly, it makes no difference with respect to this operation whether the deforming agents are physical environmental features or behavioral interactions with coupled organisms. Therefore, although the nervous system operates in a state-to-state fashion, time as a mode of behavior enters in the determination of its states through the descriptive domain as a component in the domain of behavior of the organism. The same occurs with any other component of the domain of descriptions which even though they do not represent states of the nervous system they act, as any behavior, as selectors of its path of structural change. This is so even with notions like *beauty, freedom* and *dignity* which, as descriptions arise in the domain of behavior of the organism through distinctions referred to it as a result of the coupling of the phenomenology of the nervous system as a closed neuronal network and the domain of interactions of the organism.

We have not given a formal description of the nervous system in the language of anatomy or electrophysiology because our purpose was to disclose the organization of the nervous system as a closed neuronal network, and the languages of neurophysiology and anatomy through their references to function and input and output relations imply the notion of an open system. The distinction is significant because the disclosure of the organization of the nervous system as that of a closed neuronal network leads to a fundamental notion:

The correspondence that the observer sees between the conduct of the organism and the environmental conditions with which this conduct appears to cope, reveals the structural coupling of the organism (nervous system included) to its ambience as this structural coupling is conserved through philogenic and ontogenic selection. This correspondence, therefore, does not reveal any particular feature or property of the connectivity of the nervous system that would permit it to operate with representations of the ambience in its computation of the adequate conduct of the organism.

GLOSSARY

This glossary only contains words that in this work are given a specific meaning or words that are new. All the definitions are direct quotations from the text.

ALLOPOIETIC MACHINE: machines that have as product of their functioning something different from themselves, as in a car.

AUTONOMY: the condition of subordinating all changes to the maintenance of the organization. Self-asserting capacity of living systems to maintain their identity through the active compensation of deformations.

AUTOPOIETIC MACHINE: a machine organized (defined as a unity) as a network of processes of production, transformation and destruction of components that produces the components which: (i) through their interactions and transformations regenerate and realize the network of processes (relations) that produced them; and (ii) constitute it as a concrete unity in the space in which they exist by specifying the topological domain of its realization as such a network.

AUTOPOIETIC SPACE: an autopoietic organization constitutes a closed domain of relations specified only with respect to the autopoietic organization that these relations constitute, and thus it defines a space in which it can be realized as a concrete system, a space whose dimensions are the relations of production of the components that realize it.

BIOLOGICAL EXPLANATION: a reformulation in terms of processes subordinated to autopoiesis, that is, a reformulation in the biological phenomenological domain.

BIOLOGICAL PHENOMENON: the biological phenomenology is the phenomenology of autopoietic systems in the physical space and a phenomenon is a biological phenomenon only to the extent that it depends in one way or another on the autopoiesis of one or more physical autopoietic unities.

CODING: A notion which represents the interactions of the observer, not a phenomenon operative in the observed domain. A mapping of a process that occurs in the space of autopoiesis onto a process that occurs in the space of human design (heteropoiesis) and, thus, not a reformulation of the phenomenon.

COGNITIVE DOMAIN: the domain of all the interactions in which an autopoietic system can enter without loss of identity.

COMMUNICATIVE DOMAIN: a chain of interlocked interactions such that although the conduct of each organism in each interaction is internally determined by its autopoietic organization, this conduct is for the other organism a source of compensable deformations.

COUPLING (OF UNITIES): whenever the conduct of two or more unities is such that the conduct of each one is a function of the conduct of the others.

DIVERSITY: variations in the mode in which identity is maintained.

EVOLUTION: history of change in the realization of an invariant organization embodied in independent unities sequentially generated through reproductive steps, in which the particular structural realization of each unity arises as a modification of the preceding one (or ones) which, thus, constitutes both its sequential and historical antecedent.

EXPLANATION: a reformulation of a phenomenon in such a way that its elements appear operationally connected in its generation.

FUNCTION: notion that arises in the description made by the observer of the components of a machine or system in reference to an encompassing entity, which may be the whole machine or part of it and whose states constitute the goal that the changes in the components are to bring about.

HETEROPOIESIS: the space of human design.

HISTORICAL PHENOMENON: a process of change in which each state of the successive states of a changing system arises as a modification of a previous state in a causal transformation and not *de novo* as an independent occurrence.

HOMEOSTATIC MACHINES: the condition of maintaining constant or within a limited range of values some of their variables.

INDIVIDUALITY: maintenance of identity by an autopoietic machine independently from its interactions with an observer.

LINGUISTIC DOMAIN: a consensual domain in which the coupled organisms orient each other in their internally determined behavior through interactions that have been specified during their coupled ontogenies.

MACHINE: a unity in the physical space, defined by its organization, which connotes a non-animistic outlook, and whose dynamisms is apparent.

MACHINE, PURPOSE OR AIM OF: the use to which a machine can be put by man, sometimes its product. A descriptive device to reduce the task of conveying to a listener the organization of a particular machine.

MECHANICAL PHENOMENOLOGY: the phenomenology generated by

relations between processes realized through the properties of components.

MECHANICISM: a biological outlook which asserts that the only factors operating in the organization of living systems are physical factors, and that no non-material vital organizing force is necessary.

OBSERVER: a system that through recursive interactions with its own linguistic states may always linguistically interact with its own states as if with representations of its interactions.

ONTOGENY: the history of the structural transformations of a unity.

ORGANIZATION: the relations that define a system as a unity, and determine the dynamics of interaction and transformations which it may undergo as such a unity, constitute the organization of the system.

PHENOMENOLOGICAL DOMAIN: defined by the properties of the unity or unities that constitute it, either singly or collectively through their transformations or interactions. Thus whenever a unity is defined or a class of unities is established which can undergo transformations or interactions, a phenomenological domain is defined.

PHYSICAL SPACE: the space within which the phenomenology of autopoiesis of living systems takes place.

PURPOSE: the possession of an internal project or program represented and realized through the components of a unity.

REGULATION: a notion valid in the domain of description of heteropoiesis, that reflects the simultaneous observation and description made by the designer (or its equivalent) of interdependent transitions of the system that occur in a specified order and at specified speeds.

RELATIONS OF CONSTITUTION: determine that the components produced constitute the topology in which the autopoiesis is realized.

RELATIONS OF ORDER: determine that the concatenation of the components in the relations of constitution, specification and order be the ones specified by the autopoiesis.

RELATIONS OF SPECIFICITY: determine that the components produced be the specific ones defined by their participation in the autopoiesis.

REPRODUCTION: any of the processes of replication, copying or self-reproduction.

SELECTION: a process of differential realization of a production of unities in a context that specifies the unitary organization that can be realized. In a population of autopoietic unities, selection is a process of differential realization of autopoiesis, and hence, of differential self-production.

SELF-CONSCIOUSNESS: the domain of self-observation.

SELF-REPRODUCTION: when a unity produces another with a similar organization to its own, through a process that is coupled to the process of its own specifications. Only autopoietic systems can self-reproduce.

SPECIES: a population or collection of populations of reproductively interconnected individuals which, thus, are nodes in a historical network.

STATICAL PHENOMENOLOGY: the phenomenology generated by the relations between properties of components.

STRUCTURE: the actual relations which hold between the components which integrate a concrete machine in a given space.

SYSTEM: any definable set of components.

TELEONOMY: the element of apparent purpose or possession of a project in the organization of living systems, without implying any vitalistic connotations. Frequently considered as a necessary if not sufficient definitory feature of the living organization.

UNITY: that which is distinguishable from a background, the sole condition necessary for existence in a given domain. The nature of a unity and the domain in which the unity exists are specified by the process of its distinction and determination; this is so regardless of whether this process is conceptual or physical.

BIBLIOGRAPHY

Ashby, W. Ross [1960], *Design for a Brain: The Origin of Adaptive Behavior* (John Wiley and Sons, Inc., New York), Second Edition.

Berkeley, George [1709], *An Essay Towrds a New Theory of Vision.*

Berkeley, George [1710], *Treatise Concerning the Principles of Human Understanding.*

Bernal, J. D. [1965], 'Molecular Matrices for Living Systems', in *The Origins of Prebiological Systems and of Their Molecular Matrices* (Academic Press, New York and London).

von Bertalanffy, Ludwig [1960], *Problems of Life: An Evaluation of Modern Biological and Scientific Thought* (Harper Torchbooks, New York), First Edition.

Chomsky, N. [1968], *Language and Mind* (Harcourt, Brace and World, New York).

Commoner, B. [1965], 'Biochemical, Biological and Atmospheric Evolution', *Proc. of the National Academy of Science,* **53**, pp. 1183–1194.

Gardner, R. A. and B. T. Gardner [1969], 'Teaching Sign Language to a Chimpanzee', *Science,* **165**, pp. 664–672.

Gazzaniga, M. S., J. E. Bogen, and R. W. Sperry [1965], 'Observations on Visual Perception After Disconnection of the Cerebral Hemispheres in Man', *Brain,* **88**, Part II, pp. 221–236.

Geschwind, N. [1964], 'The Development of the Brain and the Evolution of Language', Monograph Series on Languages and Linguistics, 17, C. I. J. M. Stuart (Georgetown University Press, Washington), pp. 155–169; reprinted in *Selected Papers on Language and the Brain, Boston Studies in the Philosophy of Science* **XVI**, Robert S. Cohen and Marx W. Wartofsky, eds., (D. Reidel Publishing Company, Dordrecht and Boston, 1974), pp. 86–104.

Geschwind, N. [1965], 'Disconnexion Syndromes in Animals and Man', *Brain,* **88**, pp. 237–294 and 585–644; reprinted in *Selected Papers on Language and the Brain, Boston Studies in the Philosophy of Science* **XVI**, Robert S. Cohen and Marx W. Wartofsky, eds., (D. Reidel Publishing Company, Dordrecht and Boston, 1974), pp. 105–236.

Gibson, J. J. [1950], *The Perception of the Visual World* (Allen and Unwin, London).

Held, R. and A. Hein [1963], 'Movement-Produced Stimulation in the Development of the Visual Guided Behavior', *J. Comp. and Phys. Psychol.* **56**, No. 5., pp. 872–876.

Henderson, L. J. [1913], *The Fitness of the Environment* (Macmillan, New York).

Jay, P. C. (ed.) [1968], *Primates; Studies in Adaptation and Variability* (Holt, Rinehart and Winston, New York).

Kilmer, W. L., W. S. MuCulloch and J. Blum [1968], 'Towards a Theory of Reticular Formation' in *The Mind: Biological Approaches to Its Function*, W. C. Corning and John Balaban, eds., (John Wiley and Sons, Inc., New York).

Lettvin, J. Y., H. R. Maturana, W. S. McCulloch and W. H. Pitts [1959], 'What the Frog's Eye Tells the Frog's Brain', *Proceedings of the IRE* **47**, No. 11, pp. 1940–1959.

Lilly, J. G. [1967], *The Mind of the Dolphin* (Doubleday and Company, New York).

Lindauer, M. [1967], *Communication Among Social Bees* (Harvard University Press, Cambridge, Massachusetts).

Lorenz, K. [1966], *Evolution and Modification of Behavior* (Methuen and Co., Ltd., London).

Maturana, H. R., J. Y. Lettvin, W. S. McCulloch and W. H. Pitts [1960], 'Anatomy and Physiology of Vision in the Frog (*Rana pipiens*)', *J. of Gen. Physiol.* **43**, No. 6 Part 2, pp. 129–175.

Maturana, H. R. and S. Frenk [1963], 'Directional Movement and Horizontal Edge Detectors in the Pigeon Retina', *Science* **142**, pp. 977–979.

Maturana H. R. [1965], 'Especificidad versus Ambiguedad en la Retina de los Vertebrados', *Biologica* **36**, pp. 69ff.

Maturana, H. R., G. Uribe and S. Frenk [1968], 'A Biological Theory of Relativistic Color Coding in the Primate Retina', *Arch. Biologia y Med. Exp.*, Suplemento No. 1 (Santiago, Chile).

Morrell, F. [1967], 'Electrical Signs of Sensory Coding', in *The Neurosciences, A Study Program*, Quarton, ed., (The Rockefeller University Press, New York), pp. 452–468.

Segundo, J. P. and H. D. Perkel [1969], 'The Nerve Cells as Analyzers of Spikes', in *The Interneuron*, M. A. B. Brazier, ed., (University of California Press, Berkeley and Los Angeles).

Varela, F. [1979], *Principles of Biological Autonomy*, (Elsevier-North Holland, New York).

Wittgenstein, L. [1922], *Tractatus Logico-Philosophicus* (Routledge and Kegan Paul, London).

Young, J. Z. [1967], 'On the Organization of Living Memory Systems', in *Journeys In Science: Small Steps Great Strides*, D. L. Arm, ed., (The Twelfth A. F. D. S. R. Science Seminar).

INDEX OF NAMES

BOSTON STUDIES IN THE PHILOSOPHY OF SCIENCE

Editors:
ROBERT S. COHEN and MARX W. WARTOFSKY
(Boston University)

1. Marx W. Wartofsky (ed.), *Proceedings of the Boston Colloquium for the Philosophy of Science 1961-1962.* 1963.
2. Robert S. Cohen and Marx W. Wartofsky (eds.), *In Honor of Philipp Frank.* 1965.
3. Robert S. Cohen and Marx W. Wartofsky (eds.), *Proceedings of the Boston Colloquium for the Philosophy of Science 1964-1966. In Memory of Norwood Russell Hanson.* 1967.
4. Robert S. Cohen and Marx W. Wartofsky (eds.), *Proceedings of the Boston Colloquium for the Philosophy of Science 1966-1968.* 1969.
5. Robert S. Cohen and Marx W. Wartofsky (eds.), *Proceedings of the Boston Colloquium for the Philosophy of Science 1966-1968.* 1969.
6. Robert S. Cohen and Raymond J. Seeger (eds.), *Ernst Mach: Physicist and Philosopher.* 1970.
7. Milic Capek, *Bergson and Modern Physics.* 1971.
8. Roger C. Buck and Robert S. Cohen (eds.), *PSA 1970. In Memory of Rudolf Carnap.* 1971.
9. A. A. Zinov'ev, *Foundations of the Logical Theory of Scientific Knowledge (Complex Logic).* (Revised and enlarged English edition with an appendix by G. A. Smirnov, E. A. Sidorenka, A. M. Fedina, and L. A. Bobrova.) 1973.
10. Ladislav Tondl, *Scientific Procedures.* 1973.
11. R. J. Seeger and Robert S. Cohen (eds.), *Philosophical Foundations of Science.* 1974.
12. Adolf Grünbaum, *Philosophical Problems of Space and Time.* (Second, enlarged edition.) 1973.
13. Robert S. Cohen and Marx W. Wartofsky (eds.), *Logical and Epistemological Studies in Contemporary Physics.* 1973.
14. Robert S. Cohen and Marx W. Wartofsky (eds.), *Methodological and Historical Essays in the Natural and Social Sciences. Proceedings of the Boston Colloquium for the Philosophy of Science 1969-1972.* 1974.
15. Robert S. Cohen, J. J. Stachel and Marx W. Wartofsky (eds.), *For Dirk Struik. Scientific, Historical and Political Essays in Honor of Dirk Struik.* 1974.
16. Norman Geschwind, *Selected Papers on Language and the Brain.* 1974.
18. Peter Mittelstaedt, *Philosophical Problems of Modern Physics.* 1976.
19. Henry Mehlberg, *Time, Causality, and the Quantum Theory*: Vol. 1, *Essays on the Causal Theory of Time.* 1980.
20. Kenneth F. Schaffner and Robert S. Cohen (eds.), *Proceedings of the 1972 Biennial Meeting, Philosophy of Science Association.* 1974.
21. R. S. Cohen and J. J. Stachel (eds.), *Selected Papers of Léon Rosenfeld.* 1978.
22. Milic Capek (ed.), *The Concepts of Space and Time. Their Structure and Their Development.* 1976.

23. Marjorie Grene, *The Understanding of Nature. Essays in the Philosophy of Biology.* 1974.
24. Don Ihde, *Technics and Praxis. A Philosophy of Technology.* 1978.
25. Jaakko Hintikka and Unto Remes, *The Method of Analysis. Its Geometrical Origin and Its General Significance.* 1974.
26. John Emery Murdoch and Edith Dudley Sylla, *The Cultural Context of Medieval Learning.* 1975.
27. Marjorie Grene and Everett Mendelsohn (eds.), *Topics in the Philosophy of Biology.* 1976.
28. Joseph Agassi, *Science in Flux.* 1975.
29. Jerzy J. Wiatr (ed.), *Polish Essays in the Methodology of the Social Sciences.* 1979.
32. R. S. Cohen, C. A. Hooker, A. C. Michalos, and J. W. van Evra (eds.), *PSA 1974: Proceedings of the 1974 Biennial Meeting of the Philosophy of Science Association.* 1976.
33. Gerald Holton and William Blanpied (eds.), *Science and Its Public: The Changing Relationship.* 1976.
34. Mirko Drazen Grmek, Robert S. Cohen and Guido Cimino (eds.), *On Scientific Discovery.* 1980.
35. Stefan Amsterdamski, *Between Experience and Metaphysics. Philosophical Problems of the Evolution of Science.* 1975.
36. Mihailo Marković and Gajo Petrović (eds.), *Praxis. Yugoslav Essays in the Philosophy and Methodology of the Social Sciences.* 1979.
37. Hermann von Helmholtz: *Epistemological Writings. The Paul Hertz/Moritz Schlick Centenary Edition of 1921 with Notes and Commentary by the Editors.* (Newly translated by Malcolm F. Lowe. Edited, with an Introduction and Bibliography, by Robert S. Cohen and Yehuda Elkana.) 1977.
38. R. M. Martin, *Pragmatics, Truth, and Language.* 1979.
39. R. S. Cohen, P. K. Feyerabend, and M. W. Wartofsky (eds.), *Essays in Memory of Imre Lakatos.* 1976.
42. Humberto R. Maturana and Francisco J. Varela, *Autopoiesis and Cognition. The Realization of the Living.* 1979.
43. A. Kasher (ed.), *Language in Focus: Foundations, Methods and Systems. Essays Dedicated to Yehoshua Bar-Hillel.* 1976.
46. Peter Leonidovich Kapitza, *Experiment, Theory, Practice.* 1980.
47. Maria Luisa Dalla Chiara, *Italian Studies in the Philosophy of Science.* 1980.
48. Marx W. Wartofsky, *Models: Representation and the Scientific Understanding.* 1979.
50. Yehuda Fried and Joseph Agassi, *Paranoia: A Study in Diagnosis.* 1976.
51. Kurt H. Wolff, *Surrender and Catch: Experience and Inquiry Today.* 1976.
52. Karel Kosík, *Dialectics of the Concrete.* 1976.
53. Nelson Goodman, *The Structure of Appearance.* (Third edition.) 1977.
54. Herbert A. Simon, *Models of Discovery and Other Topics in the Methods of Science.* 1977.
55. Morris Lazerowitz, *The Language of Philosophy. Freud and Wittgenstein.* 1977.
56. Thomas Nickles, *Scientific Discovery, Logic, and Rationality.* 1980.
57. Joseph Margolis, *Persons and Minds. The Prospects of Nonreductive Materialism.* 1977.

58. Gerard Radnitzky and Gunnar Andersson (eds.), *Progress and Rationality in Science*. 1978.
59. Gerard Radnitzky and Gunnar Andersson (eds.), *The Structure and Development of Science*. 1979.
60. Thomas Nickles, *Scientific Discovery: Case Studies*. 1980.
61. Maurice A. Finocchiaro, *Galileo and the Art of Reasoning*. 1980.